A WORLD OF PLANTS

A WORLD OF PLANTS

The Missouri Botanical Garden

With essays by
CHARLENE BRY, MARSHALL R. CROSBY
and PETER LOEWER

Photographs by
KIKU OBATA

HARRY N. ABRAMS, INC.
Publishers, New York

Project Director: Robert Morton
Editor: Ruth A. Peltason
Designer: Judith Michael

Page 1:
The Fountain Angel *(c. 1902) by Raffaello Romanelli is thought to represent Persephone, queen of the underworld. Originally water spouted from the four dogs' heads at the base of the figure.*

Pages 2–3:
Pink and white hyacinths and tulip 'Corsage'.

Pages 6–7:
Tulip 'Near East' and Tulipa fosteriana *'Purissima'.*

LIBRARY OF CONGRESS
CATALOGING-IN-PUBLICATION DATA

Bry, Charlene.
A world of plants : the Missouri Botanical Garden / with essays by Charlene Bry, Marshall Crosby, and Peter Loewer ; photographs by Kiku Obata.
p. c.m.
Includes index.
ISBN 0-8109-1772-6
1. Missouri Botanical Garden I. Crosby, Marshall R.
II. Loewer, H. Peter. III. Title.
QK73.U62M573 1989
580′.74′477866—dc19 89-204
CIP

Published in 1989 by Harry N. Abrams, Incorporated, New York

A TIMES MIRROR COMPANY
Printed and bound in Japan

CONTENTS

A WORLD OF PLANTS

FOREWORD

Botanical gardens are many things to many people. Each year, over 700,000 visitors come to the Missouri Botanical Garden to enjoy its horticultural displays; over 100,000 participate in education programs; and over 3,000 scientists and students use its herbarium and library collections in their research. In addition to these direct uses of the Garden's facilities and resources, many others benefit through the Garden's popular and scientific publications, public lectures, and the horticultural answer service. These three areas of activity—horticultural display, education, and research—are the kinds of activities that define museums, whether they be art museums, natural history museums, specialist museums such as transportation museums or history museums, or botanical gardens.

Like many other museums, the Missouri Botanical Garden is probably best known for its displays. In the early 1850s when the Garden's founder, Henry Shaw, conceived the idea of establishing a botanical garden in his adopted home of St. Louis, he thought first of creating beautiful displays for the people to enjoy. After opening the Garden in 1859, Shaw spent the remaining thirty years of his life making it a beautiful public garden, constructing greenhouses, and developing outdoor displays. Today we continue this tradition of indoor and outdoor display. Some of the features created by Henry Shaw remain important to the Garden's display programs, such as the Linnean House with its camelias, or his Tower Grove House and its nearby grove of sassafras and other trees. Most of the buildings and plants known to Henry Shaw have been replaced by more modern structures, newly developed horticultural varieties, or newly discovered species of wild plants. The Climatron was a pioneering venture in greenhouse construction; our rose gardens introduce and test new varieties every year; and our lovely grove of dawn redwoods are grown from seed from trees that were unknown to science until 1941, when a Chinese forester discovered them in a remote valley in China.

The Blue Passion Flower, Passiflora caerulea. *In* New Illustration of the Sexual System of Carolus von Linnaeus *by Robert John Thornton (1807).*

In a sense, all display is educational. Simply viewing the Japanese Garden tells us something about a particular culture. Visiting the restored Tower Grove House teaches us about life in years past. More directly, all the plants are labeled with their names and an indication of where they naturally grow. Many have more elaborate, interpretive labels, giving the reader information about various aspects of the plant, whether this be scientific background, the use of the plant horticulturally, or the threat to the plant in its natural habitat.

Our education programs go well beyond the casual visitor to the Garden. Late in his life, Henry Shaw established close contacts between the Garden and Washington University by endowing a school of botany there. Today the Garden continues to train graduate botanists at Washington University and other area universities. Our education and horticulture departments provide educational opportunities for preschoolers through senior citizens. The programs are offered both on the Garden's grounds and at schools throughout the area.

Like most museums, the Garden's research programs and collections are less well known to the general public than its other activities. However, they are central to our mission of increasing and disseminating knowledge about plants. We maintain approximately 30,000 horticultural accessions, but our research collection of herbarium specimens numbers over 3.5 million accessions, and there are over 110,000 volumes in our botanical library. Fully one-third of our budget is dedicated to research. These holdings, which include plants collected before the American Revolution and books printed before Columbus discovered America, are maintained and added to by our staff and other researchers around the world. Our research has always had as one of its goals accumulating information that is useful in understanding the nature of plants, their relationships to one another, and in constructing classifications that indicate these relationships. Just as importantly, this information provides the foundation for the beneficial exploitation of plants, whether for the discovery of new drugs, the improvement of food or other crops, or the development of new, beautiful horticultural forms. These kinds of activities—finding out about plants and using the resulting information to improve the lot of man—dates to the early days of the Garden. George Engelmann, Henry Shaw's botanical mentor, was interested in a wide variety of plants and carried out pioneering studies of many groups that are of great use and benefit to man. His early studies of North American pines and other conifers, grapes, and cacti form the basis for much subsequent knowledge of these plants and the development of ways to use them. Our research programs today include exploration of many areas of Latin America and Africa, and detailed studies of plants in the Americas, Africa, and Asia.

The Garden is a world of plants, a living museum devoted to their display, to educating people about them, and to research to increase our knowledge about them. When Henry Shaw died in 1889, he willed his Garden to a self-perpetuating Board of Trustees. Within days of his death, this group of St. Louisans accepted this trust and the responsibility of forever keeping and maintaining a botanical garden for the cultivation and propagation of plants, to be easily accessible to the public, with museum and library devoted to the science of botany. Generations of trustees have kept this trust and seen that facilities and programs were developed to meet the changing needs of a changing world. We look back with pride on our accomplishments over the past one hundred years and with excitement and optimism for the future.

Peter H. Raven
Director
Missouri Botanical Garden

THE BOTANICAL WORLD

Our Earth is the home of several million species, or different kinds, of living organisms. Each is distinct from all the others, exhibiting unique features or combinations of features of shape, or structure, or chemistry, or other characteristics. Plants make up only about a quarter of a million of these species of organisms, but most of the others—animals, bacteria, fungi—are ultimately completely dependent on plants as sources of food and oxygen. Plants are exploited by other organisms in many other ways, too, and clearly humans are the most creative users of plants. We use plants or plant products for building, furnishing, and decorating our shelters and for making and coloring our clothing. Plants are used as food for domesticated animals, which in turn serve as food and fiber, as beasts of burden, and as pets. Plants are important sources of fuel. Large numbers of us cultivate or gather wild plants for fuel, while others burn fossil plant products. Plants form the basis for many medicines and other useful drugs.

Because plants and many other living organisms vary greatly, it is difficult to make generalizations about their structure. Further, when discussing the numbers of various groups of plants, their distribution, and usefulness, qualifiers and round numbers are used, reflecting the incompleteness of our knowledge about many aspects of plant life.

WHAT ARE PLANTS?

Plants are members of the plant kingdom, one of the five fundamental groups of living organisms now recognized. (The other kingdoms are, broadly, bacteria, algae, fungi, and animals.) To make sense of this, a definition of what characterizes a plant is needed. First, just consider these few familiar and perhaps not so familiar plants: trees, bushes, and vines are plants, but these terms describe the *habits* that some plants

Sacred lotus, Nelumbo nucifera. *In* New Illustration of the Sexual System of Carolus von Linnaeus *by Robert John Thornton (1807).*

Slash pine, Pinus elliottii. Revision of the genus Pinus *in* Transactions of the Academy Science of St. Louis, *vol. 4 by George Engelmann (1880). Pines and other conifers bear their seeds, shown in figures 29 and 30, on scales, figures 26 and 27, which are clustered into cones, 21 through 25.*

Opposite:
Orange daylily, Hemerocallis fulva. *In* Botanical Magazine, *vol. 2 by William Curtis (1788). The term* daylily *refers to a small group or genus of plants known botanically as* Hemerocallis, *which means beautiful for a day, referring to the fact that individual flowers of this plant last only one day. The seven species of* Hemerocallis *are all native to Asia, where the flowers, stems, and roots have been used for centuries as sources of food and flavorings. The thousands of horticultural varieties of daylilies have been derived from these species.*

assume, that is to say, the forms they take in growing; flowers, fruits, and vegetables are also terms that refer to *parts* of plants rather than to kinds of plants. Examples of some kinds of plants are pines, palms, and oaks; most of these take the form of trees, but each has certain characteristics that make it distinct from all other plants. Thus, pines produce their seeds in cones of a particular kind and have their leaves arranged in distinctive groups. Privets, roses, and honeysuckles are also different kinds of plants, and there are many species of each. Marigolds, lilies, orchids, asparagus, oranges, apples, and pineapples are other familiar plants. Less familiar are cycads, ferns, horsetails, quillworts, mosses, liverworts, and hornworts.

Even this short list of plants brings to mind a great diversity of sizes, shapes, colors, and structures. Still, there are fundamental features of plants that differentiate them from all other living organisms. The two most important features are that plants are photosynthetic and that they produce embryos. Being photosynthetic means containing certain pigments, chlorophylls being the best known, that absorb light energy and convert it to chemical energy. This process is known as *photosynthesis.*

Plants are not the only photosynthetic organisms: many algae and some bacteria are photosynthetic, but they do not produce embryos. Embryos are young organisms that have resulted from the union of sexual cells, an egg and a sperm, and that undergo early growth while remaining attached to and surrounded by specialized tissue of their mother; they also receive their nourishment from their mother. Although plants and many animals each produce embryos, a main distinction is that animals are not photosynthetic.

The Distribution of Plants

Of the quarter million species of plants on the Earth, a little less than one-third occur in temperate areas and the remaining species, about 180,000, in the tropics. Even within the tropics the diversity of species is uneven and remarkable. By far the richest tropical area is the New World tropics, the area extending just north of Cuba and south near the latitude of Rio de Janeiro in South America. Here there are about 90,000 species of plants, about one-third of all plants. Tropical Africa has about 35,000 species on the mainland with an additional 8,500 on the nearly totally tropical island of Madagascar, located off the southeastern coast. Tropical Asia has about 40,000 species.

In the three northwestern Andean countries of Colombia, Ecuador, and Peru alone there are about 45,000 species of plants, nearly one-fifth of all the world's flora. These three countries have a combined area of just over one million square miles. Europe has four times this area; its flora is about 12,000 species, just under one-fourth that of these three tropical countries. The United States east of the Mississippi River has an area just under 900,000 square miles, about the same as these three countries. Its flora is about 10,000 species, less than one-fourth that of

N 64
Publish'd by W. Curtis, Botanic Garden, Lambeth Marsh.

A Puerto Rican calabash, Crescentia portoricensis, *growing in the Research Greenhouse. This species of Bignoniaceae or trumpet creeper family occurs naturally in two areas of Puerto Rico, where its existence is threatened because of the degradation of its habitat. Only three or four plants are known in the wild. Garden botanist Alwyn Gentry collected seeds and established the species in the Garden's Research Greenhouse, where about ten plants now thrive. Plants have also been distributed to other botanical gardens, including one in Hawaii, where the species may grow out-of-doors.*

the tropical countries. More broadly, 20,000 species occur in North America, north of Mexico.

Although there is no comprehensive list of the plants of the tropical New World, there has been a great deal of research in Panama, an area of about 30,000 square miles. Its flora has been studied intensively but not exhaustively by botanists from the Missouri Botanical Garden and other institutions over the past sixty years. Beginning in 1943 and ending in 1982, the Garden published an illustrated and descriptive *Flora of Panama* in twelve volumes. It enumerated about 6,500 species. Continued exploration of Panama was uncovering many new species even as the *Flora* was being produced, and in order to bring the list of species up-to-date, a two-volume checklist—giving the name of each plant and some information about its local occurrence but no descriptions of the plants known from Panama—was published in 1987. The checklist enumerated 8,500 species, an increase of 2,000 species, nearly one-third, over the number reported in the *Flora,* finished just five years earlier.

The state of Missouri has an area of 70,000 square miles, about two and one-half times that of Panama. The *Flora of Missouri,* published in 1964 and authored by the late Missouri Botanical Garden botanist Julian A. Steyermark, listed about 2,400 species for the state, about one-third the number now known from the much smaller Panama. In the twenty-five years since the *Flora of Missouri* was published, continued collecting and study have added only about 200 species to the state's known flora, an increase of less than ten percent.

Examples like these could be repeated many times, demonstrating time and again the relative richness of tropical areas compared to temperate areas and the lack of knowledge about the plants of tropical areas compared to temperate areas. The richness of the tropics is an interesting fact, and scientists continue to make and test hypotheses to explain it. The demonstrated lack of knowledge about tropical plants argues for additional studies to find out more about them. Such studies add to our general knowledge about the diversity, distribution, and relationships of plants. With additional basic knowledge would surely come the discovery of new plant products from the tropics and ways to effectively manage tropical areas for their maximum benefit to mankind.

The sheer numbers of plants that occur in the New World are overwhelming. It is noteworthy that none of these were known to European man five hundred years ago, before the discovery of America by Columbus in 1492. An idea of the importance of New World plants may be gained by a few examples of a few of them.

Corn, *Zea mays,* is a New World plant; it is the third most important food crop on Earth, being out-produced only by wheat and rice.

Potatoes, *Solanum tuberosum,* and tomatoes, *Lycopersicon esculentum,* both members of the same family, Solanaceae, are other well-known New World foods. They are the first- and second-ranking vegetables in terms of food production. Other members of the family include cayenne and chili peppers, paprika, and bell peppers (all members of the genus

Capsicum), tobacco, *Nicotiana,* used as a drug and sometimes cultivated as an ornamental, and petunia, *Petunia,* a widely used ornamental.

The pineapple, *Ananas comosus,* is a member of the bromeliad family, Bromeliaceae, a family that is endemic, or naturally confined, to the New World. In addition to the pineapple, the family contains about 2,000 species, some of which are sources of fiber and some of which are cultivated as ornamentals because of their showy leaves or flower bracts. The pineapple was first domesticated in South America and is now grown throughout the tropics as a source of vitamin-rich fruit and juice.

Another exclusively New World family is the cactus family, Cactaceae. These curious plants are widely cultivated as ornamentals, but certain prickly pears, the genus *Opuntia,* are grown for their edible fruits and as cattle-enclosing fence rows.

Although rubber may be produced from a wide variety of plants in several families, virtually all of the world's natural rubber comes from *Hevea brasiliensis.*

Cultivated ornamentals, all members of the sunflower family, include *Dahlia, Tagetes* (marigold), *Zinnia,* and, of course, *Helianthus,* sunflower itself. The annual sunflower, *Helianthus annuus,* is an important source of vegetable oil, especially in the Soviet Union, and the Jerusalem artichoke is also a member of this genus.

Yucca and *Agave,* both members of the same family, Agavaceae, are widely cultivated ornamentals; tequila is distilled from the sap of the latter, and fibers from its leaves produce sisal.

The orchid genus *Vanilla* contains about one hundred species from the New World. One of these, *Vanilla fragrans,* is the source of the flavoring vanilla, produced from the fermentation of its seed pods. It is now cultivated mainly in Madagascar, which produces about three-quarters of the world's vanilla.

Most of the New World plants listed above are tropical or subtropical in origin. In fact, North America has provided remarkably few plants that have been exploited on a large scale. The annual sunflower, though, is a good example of a North American plant of wide use. But in spite of the fact that the plants of the New World have been collected and studied with some degree of organization and vigor for several hundred years now, we still know remarkably little about most of these plants. For North America there is still no single reference to which one can turn for even basic information about the plants of this region, though the Missouri Botanical Garden, in cooperation with other U.S. and Canadian institutions, has recently launched a renewed effort to produce such a work. Computerization of much of the information will greatly aid this project, but it will still be the end of the century before it is finished. By computerizing the information—descriptions, specimen information, names, ecological data—the production of printed volumes will be expedited, with the added benefit of the information being more easily accessible.

The potential of computerized information and up-to-date lists on

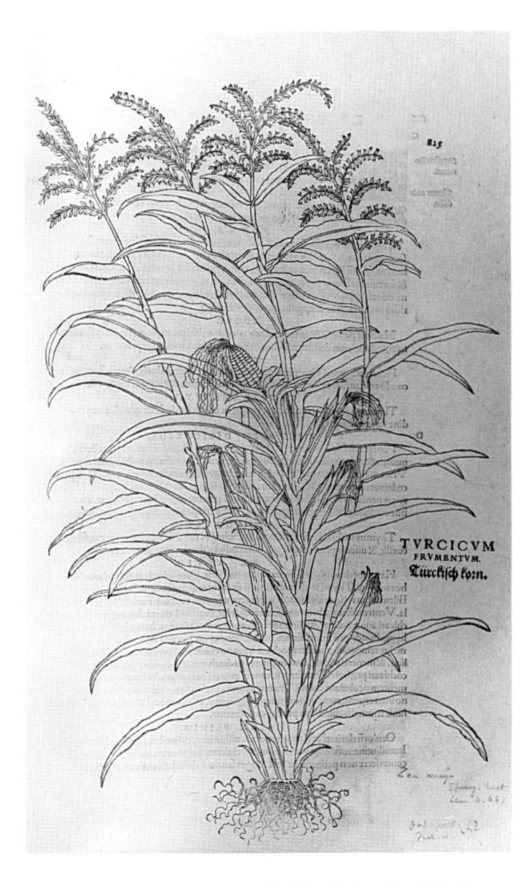

Corn, Zea mays. *In* De Historia Stirpium *by Leonhard Fuchs (1542). Corn is the most important economic plant of New World origin. It formed a dietary staple for Amerindians and was widely cultivated even before the coming of Europeans in 1492. Today, it is grown worldwide.*

Overleaf, left:
Pineapple, Ananas comosus. *In* Uitgezochte Planten *by Christophor Jacob Trew (1771–1773).*

Overleaf, right:
Queen-of-the-night, Selenicereus grandiflorus. *In* New Illustration of the Sexual System of Carolus von Linnaeus *by Robert John Thornton (1807).*

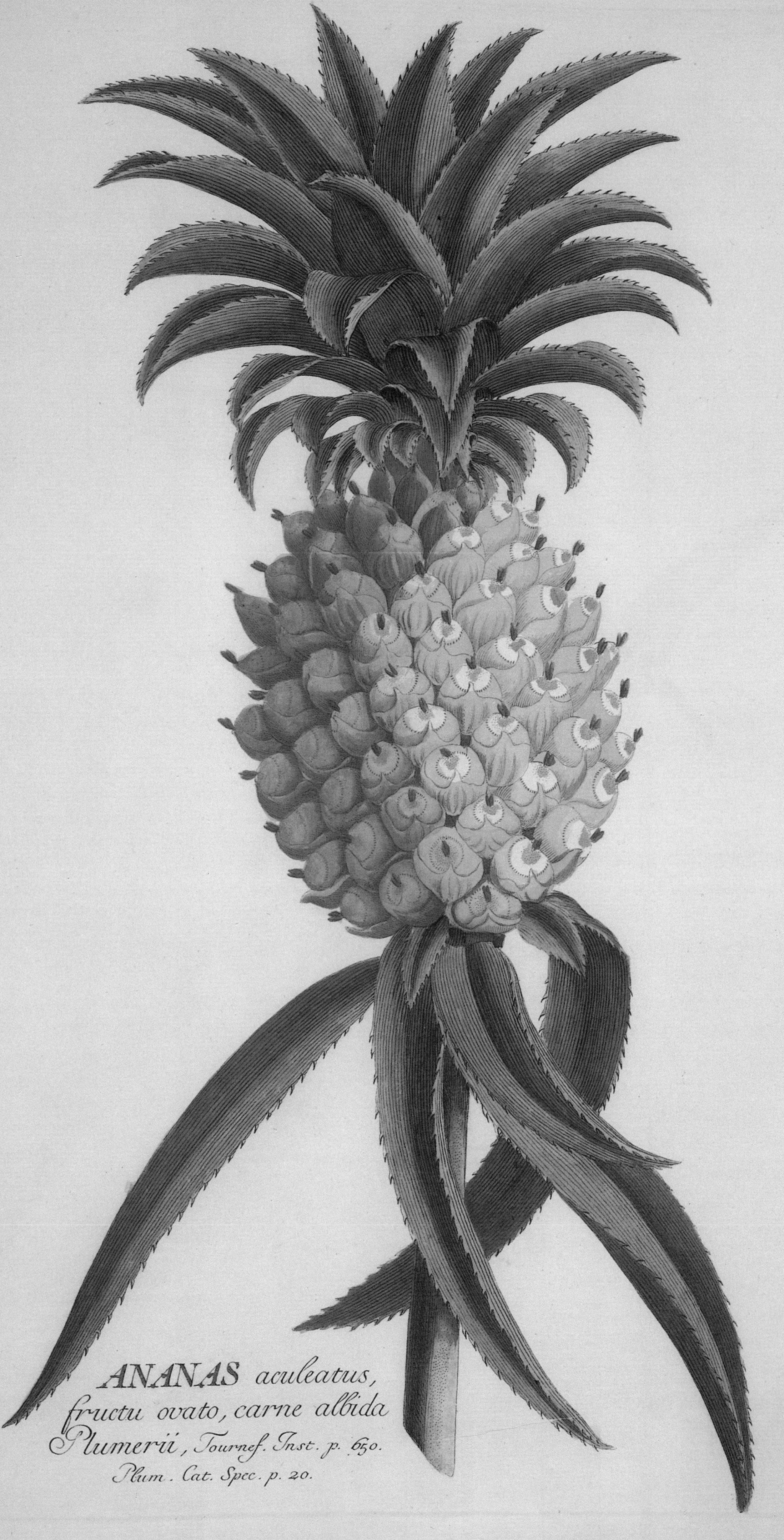

ANANAS aculeatus,
fructu ovato, carne albida
Plumerii, Tournef. Inst. p. 650.
Plum. Cat. Spec. p. 20.

This tropical rain forest in Brazil was destroyed in an attempt to replace it with fast-growing trees for pulp wood. Several factors, including economic conditions and a lack of understanding about beneficial land management, doomed the project, and it was abandoned.

plants in the tropics would be immensely valuable. However, given the rapidity with which new species are still being found there, it might be argued that it is still too early to attempt such a listing. But the compilation of such lists is just one stage in the process of developing an overall knowledge of plants. They provide guides to identification, highlight groups of plants that need further study, and form the foundations of more detailed, descriptive floras of areas.

Today there is an even greater urgency to learning about the tropics. Tropical forests are literally disappearing due to a variety of reasons, and this destruction will certainly have wide-ranging adverse effects.

On a world scale, tropical forests are being cleared at the rate of about 20,000 square miles a year for lumber. Returning to the example of Panama, this means that every year and a half, an area the size of that country is being cleared.

Shifting or slash-and-burn agriculture causes even greater disruption of tropical forests. This practice involves the cutting of forests and burning of trees and other cut materials in order to release the nutrients stored in them: in tropical areas most nutrients are in the living matter, or biomass, rather than in the soils, as in temperate areas. After a few years of cultivation, the soils become exhausted, and the slash-and-burn agriculturist must move to a new forest area and begin the process again.

Other factors involved with the destruction of tropical forests are cutting for firewood and clearing for cattle raising—it is marginally cheaper to raise cattle and ship it to temperate areas for sale rather than to raise and sell it there.

Taken together, these activities are destroying about 80,000 square miles each year. Given this steady pressure the forests cannot last much longer. There are about two million square miles of tropical forests remaining on Earth, down from about twice that amount at the dawn of agriculture, about 10,000 years ago when humans began deliberately clearing ground for cultivation. But at current rates, these forests will only last about another thirty years, or about one generation. But this kind of statistical projection is misleading because conditions there are escalating—including the size of the population. About half the world's present population of five billion lives in the tropics. In another thirty years, the population of the Earth will have increased to about eight billion, and the proportion of those living in the tropics will increase to about two-thirds, meaning that there will be about five billion people living in the tropics—or about the same number as are on the whole planet today. This kind of tremendous increase in an area decreasing in available, livable space portends serious problems worldwide.

THE NAMING OF PLANTS

In the list of examples of New World plants, two sorts of names were used to designate them. Corn, potatoes, and rubber are familiar enough, but the names that follow them may seem foreign. These are the scientific or botanical names of the plants, usually derived from or based on Latin words and printed in *italic* for emphasis. The familiar names may be ambiguous or may vary. Many English-speaking cultures use the word *corn* to mean corn, maize, and wheat. Other languages use entirely different words: maiz, mais, kidiaansch koorn, kornetas, granoturco, zboze kukurydza, mielies. In order to communicate, verbally or in printed form, across cultures and languages, a simple, unambiguous, stable, and standardized system of naming plants has been developed by scientists.

The Swedish naturalist Carl Linnaeus (1707–1778) established such a method of naming plants. The names themselves consist of two Latin or Latin-like words. This is called the binomial method of nomenclature, since it employs two words. Although the binomial method had been developed and used sporadically earlier, it was with the publication in 1753 of Linnaeus's *Species Plantarum* (which means "species of plants" in Latin) that this system became accepted and standardized throughout the small community of people then describing, classifying, and naming plants. Previous systems of naming had not been simple, whereby descriptive phrases were used to distinguish plants. Nor were they standardized; different people used different phrases for the same plant, which made for obvious confusion. Lacking standardization, there was ambiguity as to which plant a person was referring. Linnaeus used a different combination of two words for each of the approximately 7,000 known species of plants; this way each plant had a unique name, and thus avoided previous problems of ambiguity.

A practical example of the binomial method would be the flowering dogwood of eastern North America, *Cornus florida.* The first word is always capitalized and is the name of the *genus* (plural genera), or general group of plants. *Cornus* is the genus of dogwoods, and it may stand alone, if one is referring to this entire group of plants.

The second word is the name of the *species,* often called the specific name or epithet, and is not capitalized, even when it is based on a proper name. The specific name may not be used alone as it makes no sense: the same specific name may be used in conjunction with many different generic names. To refer to a plant as *florida* does not indicate to which of several genera the plant belongs.

This system of naming is similar to that used in many human cultures, except that we usually reverse the order of the names. Sam Jones is a particular person, and to speak about the Joneses means one is speaking about a general group of people. To speak about Sam is, broadly, not particularly informative, though, since there may be a Sam Jones, Sam Smith, Sam Adams, and so on.

A third italicized name, preceded by a qualifier such as *var.* (for variety) or *subsp.* (for subspecies), often follows the species name. This indicates that a morphological or geographical variant of the species is recognized. The qualifier, indicating the rank, must be used in botanical nomenclature. The abbreviation *fo.*, for form, is frequently encountered, too, usually indicating the presence of a minor, often color-based, variation of a species. Thus a pink-flowered version of the eastern dogwood might be named *Cornus florida* fo. *rubra.*

Botanical names, being Latin or Latin-like, are often said to be difficult to pronounce or understand, but familiarity soon makes them accessible. Further, many botanical names of plants are the same ones by which they are commonly known. Geranium, philodendron, magnolia, campanula, petunia, zinnia, dahlia, caladium, lantana, and cactus are each commonly used and easily pronounced. All are the correct scientific names of genera of plants. The only difference between their "common" usage and "scientific" usage is in the way they are printed: italic type and capital letters at the beginning in scientific usage. Dozens more could be listed.

Occasionally, everyday usage of Latin names for plants causes confusion. For example, the genus *Geranium* (uppercase "G," italic) and the popular houseplant called geranium (lower case "g," roman) are not the same, although they are related, both being members of the same botanical family. The houseplant is known botanically as *Pelargonium.* Printing botanical names in italic serves to distinguish them from common names that are spelled the same way but that often refer to different plants.

Genera of plants may be defined as groups of related species. Another grouping often referred to is the *family,* which may be thought of as a group of related genera. Families are also given Latin names, and they have the ending -aceae, pronounced phonetically as a-c-e. The ending is joined to the stem of the name of a genus that is included in the family. Thus, the family in which *Geranium* and *Pelargonium* are placed is called the Geraniaceae. This ending is similar to the English suffix -ous, meaning *like:* the Geraniaceae contains plants like *Geranium.*

All this explanation is a way of showing how long and to what extent effort has been made toward establishing methods of precise classifications and names. However, it remains virtually impossible to completely stabilize names for a variety of reasons. For example, concepts of the nature of species change with increasing general knowledge of the nature of life: in Linnaeus's time it was generally thought that species were the result of a single creative act; they are now generally thought to have arisen many times through evolution. Further, with the accumulation of more and more specimens for study, the morphological, geographical, and temporal variation of species becomes better understood. This may result in the detection of many species where previously only one had been recognized. Or, it may result in the recognition of fewer species, as what was once thought of sharp differences between species become recognized as only the extremes of a broad spectrum of variation.

A marble bust of Carl Linnaeus, by Howard Kretschmar, sits above the pediment over the Linnean House.

But many steps have been taken to help with the stabilization of names. Most fundamentally, Linnaeus's *Species Plantarum* has been agreed upon as the starting point for the names of many groups of

DEDICATED
TO
LINNÆUS
THE PRINCE OF
NATURE
BY Dr. GEO. ENGELMANN
ED. H. S. 1881

plants, since it did provide unique names for all of the plants known at that time. Names published before 1753, by Linnaeus or anyone else, are referred to as pre-Linnaean, and do not exist as far as naming plants is concerned. But the pre-Linnaean literature remains important to modern studies of plants because Linnaeus and his immediate successors often referred to it when publishing information about plants; thus it is consulted in order to properly interpret and understand references in subsequent publications.

Linnaeus himself set out a series of rules for naming plants, some of which were aimed at stability. These have been refined over the last nearly quarter millennium into a set of rather complex rules, known as the *International Code of Botanical Nomenclature*. Fundamentally, these state that each plant may have only one correct name, that each name must be different, and that if names compete for use, the name that was published first is the one to be used. To return to the example of dogwood, the species in eastern North America is known as *Cornus florida,* and the western North American dogwood is known as *Cornus nuttallii*. If for some reason it was determined that these were in fact the same species, a single name would have to be chosen. The *Code* provides rules for determining this name. In the case of competing names, *Cornus florida* is older, dating from 1753 versus 1840 for *Cornus nuttallii,* so it would be used for the combined species.

Knowledge of the nature of the various groups of plants had not advanced sufficiently in 1753 to distinguish even some of the major groups, mostly because studies of microscopic structure were not far advanced and because the life histories (or life cycles) of many plants were not well understood. This resulted in Linnaeus's placing unrelated plants in the same group. For example, in the genus he called *Mnium,* an ancient Greek word for moss, Linnaeus included species of both mosses and liverworts, plants now known to be very different.

Having decided that a given name can apply to only one plant in order to assure unambiguous communication, scientists had to resolve these mixtures that Linnaeus made, since his *Species Plantarum* has been the starting point for naming plants. As these problems became clearer and as the rules for naming plants became more refined, it was decided to start the nomenclature of some groups with publications other than Linnaeus, usually with works that exhibited a clear understanding of the group of plants under consideration. These are all post-1753 publications employing binomial nomenclature. Thus, botanists have designated Johannes Hedwig's *Species Muscorum,* meaning "species of mosses," published in 1801, as the starting point for most mosses. Hedwig, who died in 1799, had undertaken careful studies of mosses and similar plants during the last quarter of the eighteenth century. These studies resulted in a correct understanding of the various structures of mosses and their relationship to the life cycles of these plants. A series of beautifully illustrated books were published that made these findings available, and his posthumous *Species Muscorum* enumerated all the mosses known to him.

Shaw's agave, Agave shawii, *in the Desert House. This native of northern Baja California, Mexico, and southern California, was named for the Garden's founder, Henry Shaw, by his botanical advisor, George Engelmann.*

PUBLISHING AND INDEXING NAMES

Since the age of Linnaeus our knowledge has increased to the point where we know at least something about nearly 250,000 species of plants, an increase of slightly more than 1,000 species each year since 1753. This kind of growth also has made the application of names more complex. Each species has been assigned a Latin binomial (and in fact many have been assigned several names). The assignment of the name is governed by the rules detailed in the *International Code:* names must be published, in the traditional sense of ink on paper, in books or journals that are generally available to the botanical community; they must be accompanied by a Latin description or short diagnosis, stating what is distinctive about the species or how it differs from previously described ones; and a single specimen to which the name is permanently associated must be listed.

Publication in newspapers or oral descriptions is not allowed, for the obvious reasons that they are ephemeral. Publication in books or journals generally assures that the description will be permanently and widely available. Latin is required, because it remains a language that is reasonably widely known in the scientific community, and it has a stable grammar and vocabulary. Further, botanists would never be able to agree on a single modern language to use in its stead. This requirement is not as horrendous as it might seem: "botanical Latin" has a fairly simple grammar, a small vocabulary, and is summarized in several modern texts.

The requirement that a name be attached to an actual specimen, called the type, is important for stability. Written descriptions may be ambiguous when published or may become ambiguous as additional similar plants are discovered in the future. By associating the name with a specimen this ambiguity is removed; the specimen may always be examined to determine the correct application of the name.

Although our current *Code of Nomenclature* is fairly precise about what is required for establishing a name today, many of its provisions are not retroactive. Older names were published using broader rules, and the interpretation of the status of these names is often difficult.

As it stands, there are many more names for plants than there are plants. Probably the most common reason is that two, or often more, people will independently describe and name the same species of plant, working perhaps in different countries or states and at different times. Since species may be described in many different journal articles and books, it is virtually impossible to keep up with everything that is being or has been published. Sometimes it takes years for botanists to determine that two descriptions apply to the same species. When this happens, the name that was published earlier is taken up for the plant and the later name or names is relegated to its synonymy.

In many cases the decision as to whether two names apply to the same species is subjective, but there is one situation in which it is unequivocal. This is when the different names were published using the same plant specimen as the basis for the name. This happens more often than one might at first expect, because specimens often exist in duplicate. Thus if one were collecting an oak tree, several specimens could be made from the same tree at the same time and be distributed to several institutions. This might result in one or more botanists' describing new species, with different names and probably different dates of publication, based on the very same plant. Only one of these names—again, the earliest published—may be used for the species.

Since there have never been rules governing which journals and books could serve as places for the publication of names, names have been published in literally thousands of different places. In order to keep track of these names, various indexes are compiled, citing the name, its place of publication, and in some cases other basic information, for example, information about the type specimens.

These indexes are each massive, multivolume works, containing the bare minimum about plants, their names, species after species, and little more. An analysis of these indexes indicates that there are well over one million names for the quarter of a million plants on Earth. A handy digest for lay readers and scientists is *The Plant Book* by David Maberly, listing all the genera and families of seed plants and ferns recognized as distinct by botanists as of about 1985. *The Plant Book* also provides general information about how plants are used and which ones are of horticultural importance. The book is based on an exhaustive survey of the published literature, taking into account the latest research on each group. Of course, for many groups the latest research may be decades old: there are about 20,000 genera of seed plants and ferns. Each entry in this "dictionary" of plants is a thumbnail sketch of the makeup and distribution of the particular group listed. For each genus there is an

Umbrella tree, Magnolia tripetala. *In* Plantae Selectae *by Christophor Jacob Trew (1750–1773). Details of this illustration of an eastern North American* Magnolia *show the fundamental characteristic that defines flowering plants, the carpel, which encloses the ovules within sterile tissue, figures q, u, and t, to the left of the explanatory legend. Mature seeds, which develop from the ovules, hang from the fruit at the top of the plate, while nearly mature ones are exposed within their carpels.*

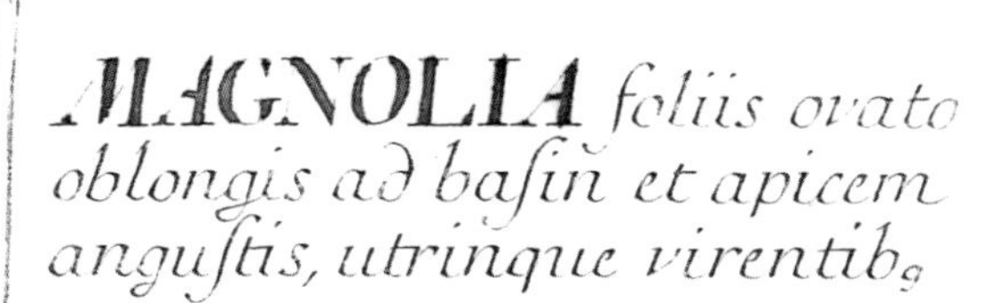

MAGNOLIA *foliis ovato oblongis ad basin et apicem angustis, utrinque virentib9*

1.1.2. gemmæ involucrum, 3.3.3.3.3. duo puncta seu glandulæ, 4. gem̄æ involucrum adhærens, 5. idem solutum et delapsum, 6. gemma nondum aperta, 7. alabastrus, 8. spatha, 9.9. ejs adhæsionis vestigium, 10.10. puncta alba pedunculi, 11. alabastrus expansioni proximus, 12.12.12. tria petala exteriora, a. ovarium, b.b.b.b. stamina, c.c.c.c. c.c. pistilla, d. stamen a facie externa, e. interna, f.f. antheræ, g. ovarium verticaliter dissectum, h.h. pistillorū axis, i.i. staminum vestigia, k.k. duo pistilla separata, l.l. eorum styli, m.m. germina, n. germen transverse o. verticaliter dissectum, p. seminum rudimenta, q.r.r.s.t.u.x. eadem partes in magnitudine aucta, y. flos, z. strobilus.

Hæc characteris explicatio conferenda est cum illa p. 31. col. a. et secundum hanc emendanda expositio p. 30. b.

estimate of the number of species, and for each family there is an estimate of the number of genera and species. If one adds up all the species, the total is 236,786, based on figures published in Maberly's book. Thus we have a total for the number of species "on the books," but new species continue to be discovered, described, and named.

One would imagine that the plants of the United States are reasonably well known. The *Gray Herbarium Index,* compiled at Harvard University, is the only index that attempts to maintain an up-to-date listing of names and other basic information about plants being described from the New World. Just as with government and other reports, the indexes that list newly published names for plants lag behind the present a bit. The most recent period for which the *Gray Herbarium Index* is up-to-date is 1986. During that year 33 species were described from the contiguous forty-eight United States. Most of these were from western states, including California, and Florida. Some were based on specimens of species that had never before been collected, while others resulted from careful studies of previously known specimens. In addition to these 33 new species, another dozen new subspecies and ten new varieties were described during the same period.

For the world as a whole, literally hundreds of new species are described each year. The staff of the Missouri Botanical Garden alone described 236 new species of plants during 1987. Most of these were from tropical areas in Latin America and Africa, reflecting the research concentrations of the Garden, and most were based on recently collected specimens, made either by the Garden's staff or sent to staff specialists from workers at other institutions.

DISTANT RELATIVES OF PLANTS

Two hundred years ago naturalists believed the natural world consisted of three kingdoms: mineral, animal, and plant. Minerals are nonliving objects, while animals and plants are living, or biological, parts of the world. In general terms, animals were organisms that did not manufacture their own food, had cells without rigid walls, and often moved. Plants were defined as usually manufacturing their own food, having cells with rigid walls, and usually being stationary.

As knowledge about organisms increased, the two-kingdom system became increasingly unsatisfactory. On a general level there were too many organisms which did not fit conveniently into either kingdom. Some "plants" did not manufacture food, such as the fungi. Some single-celled "algae" did not have rigid cell walls and were motile, though they did manufacture food. These "algae" were identical with some "animals," except for the fact that the "animals" did not manufacture food.

Today the biological world is recognized as consisting of five kingdoms, based on detailed information about the structure and biology of organisms. Animals are one kingdom; organisms traditionally thought of as plants are now placed in four kingdoms: bacteria and their relatives, algae and their relatives, fungi, and plants.

Although living organisms are no longer thought of as being either animal or plant, our increased understanding of their cellular structure and particularly the organization of the nucleus still reveals two fundamental groups, based on the degree of organization of the nucleus, that part of the cell in which the genetic material, deoxyribonucleic acid, almost always referred to simply as DNA, is located.

In one group the DNA consists of a single, circular molecule, which is not surrounded by a membrane. This simple form of nuclear organization is called *prokaryotic,* meaning pre-nucleate, implying the lack of a true nucleus. Organisms with this kind of nuclear organization form a kingdom called Monera, and most of its members are the bacteria. These organisms were formerly included in the plant kingdom, because most of them have rigid cell walls. Some contain chlorophyll and manufacture their own food. These are the blue-green "algae," now often referred to as cyanobacteria, meaning blue-green bacteria.

There are only about 2,500 species of bacteria, about one percent as many species as of plants. Although they lack size and diversity, they are enormously abundant as individuals. In fact, they are by far the most abundant kinds of organisms on Earth and play vital roles in the economy of man and of nature. They produce many substances useful to humans, and continuing research holds promise for many new products from this group of organisms. For instance, the recently developed treatment for river blindness, a serious disease in Africa spread by flies, is based on a product of bacterial fermentation. Bacteria also produce disease-curing antibiotics: neomycin, streptomycin, tetracycline, among many others. Many are involved with fermentation processes, especially of milk products, that are important to humans: yogurt, for example, is produced by bacterial activity in milk.

Other bacteria are not so welcome, and cause many serious diseases afflicting plants, humans, and other animals. Some of these include various diphtheria, pneumonia, tuberculosis, typhoid fever, and Legionnaires' disease.

Bacteria are fundamentally unicellular organisms, though in many the cells adhere together to form large masses. Sexual reproduction, involving the fusion of sexual cells or gametes and the later occurrence of reduction division or meiosis, does not occur in bacteria. Reproduction is by fission, involving an increase in cell size and the replication of the DNA molecule. At a certain point the cell simply splits, with one DNA molecule ending up in each daughter cell.

In all other organisms the DNA is contained within a membrane-bound sack, forming a distinct, organized nucleus. Furthermore, the DNA occurs in complex structures called chromosomes, through association with proteins. This form of cellular organization is referred to as *eukaryotic,* meaning truly nucleate, or having a real nucleus. Unlike the bacteria, most other organisms exhibit multicellular organization and reproduce sexually.

There are four kingdoms of these truly nucleate organisms: animals, plants, fungi, and protists (algae and their relatives).

The protists, or kingdom Protista, contain a diverse group of organisms, some of which have been considered plants. These include the algae and various kinds of molds. Also included in this kingdom are the protozoans, formerly treated as single-celled animals.

Algae is a catchall term for various unrelated single- and multi-celled organisms. The most familiar algae are the seaweeds, often quite large, multicellular organisms belonging to three quite distinct groups, or divisions, named for their predominant color, caused by pigments that mask the green of their chlorophyll. In addition to these differences, there are striking differences in the kinds of chemicals found as food reserves and in the cell walls of these organisms.

The brown algae, or Phaeophyta, are mostly marine organisms of shallow, rocky temperate oceans. Overall, they number about 1,500 species. Their generally brownish color is caused by the presence of pigments called xanthophylls. Some brown algae are huge, with kelps reaching 60 meters in length. The Sargasso Sea area of the warm north Atlantic Ocean is named for the brown alga *Sargassum.* Brown algae are sources of iodine, potassium, sodium, vitamins, and food. Kelps are particularly versatile, being used as sources of minerals, food, and industrial chemicals.

Red algae, Rhodophyta, are also predominantly marine. Their reddish color is caused by the presence of phycoerythrins. There are about 4,000 species of red algae. A few are commercially important sources of food; "Irish moss" in North Atlantic areas, and "nori" in Japan, are both derived from red algae. The widely used colloidal substance agar is derived from several kinds of red algae. Perhaps best known as the base for culture media for bacteria and other microorganisms, agar is also used to contain medicine in capsular form, and in cosmetics, desserts, and jellies. Carrageenan, used as a stabilizer in cosmetics, dairy products, and paints, is derived from another red alga. The marine Rhodophyta that are the sources of these substances are cultivated on large scales.

Green algae, or Chlorophyta, named for their predominantly bright green color caused by the presence of chlorophylls, are the most abundant algae in terms of species, numbering about 7,000. Most species occur in fresh water, though some are marine. Nonaquatic green algae occur in the soil or on tree trunks. Others are associated with entirely unrelated organisms in symbiotic relationships, most notably and abundantly with various fungi to produce lichens.

Green algae, less important to humans as other algae, are nonetheless of great interest and significance because they share several characteristics with plants. Like green algae, plants' chloroplasts contain two kinds of chlorophyll, called simply *a* and *b,* and like them, store starch in their chloroplasts. The other algae groups contain different combinations of chlorophylls, also designated by letters, and while some of these algae store starch, it is not stored within the chloroplast. Cellulose is the primary constituent of the cell walls of plants and of some Chlorophyta. For this reason they are thought to be the group from which plants originated.

Fungi, long included in the plant kingdom, are now placed in a kingdom of their own, because they exhibit so many unique characteristics. The scientific name for this kingdom is simply Fungi. Like plants, many fungi characteristically have cells with rigid walls. But the cell walls of these fungi consists of chitin rather than the cellulose found in the cell walls of plants.

There are many other fundamental differences between fungi and plants and the other kingdoms of living things. Fungi do have motile cells, found in many plant groups. They store energy in glycogen, rather than as starch as do plants. Unlike most plants, fungi do not manufacture food, being either saprophytic, deriving their nutrition from dead organisms, usually plants or animals, or parasitic, living on other living organisms.

The kingdom Fungi is divided into several divisions. Probably the most familiar are the basidiomycetes, or club fungi, so-named because of their unique club-shaped basidium, on which the sexually produced spores of these organisms are produced. Club fungi include mushrooms, usually thought of as edible, and toadstools, usually thought of as inedible or even poisonous, as well as shelf fungi, coral fungi, puffballs, earth stars, and stink horns. These names actually refer to the spore-producing portion of these fungi. Most of the fungal organism consists of a usually colorless mass of filaments known as a mycelium, which penetrates and infiltrates the substrate from which the fungus derives its nourishment.

The club fungi, of which there are about 25,000 known species, also include rusts and smuts, which are important parasites of plants. Rusts in particular are pests of many major cereal crops, including wheat, oats, and barley. Each year they cause billions of dollars worth of destruction of these and other crops around the world.

Another major group of fungi are the ascomycetes or sac fungi, named for the saclike structure in which their sexual spores are produced. There are about 30,000 known species of sac fungi, and based on the known diversity of the group and the rate at which new species are being found, described, and named, many thousands more await discovery. The most highly prized edible fungi in the world are sac fungi: truffles and morels. Yeasts are ascomycetes, used in a variety of processes, including brewing and baking. However, not all sac fungi are so delicious. Many sac fungi are food-destroying molds, others cause Dutch elm disease and Chestnut blight.

Lichens are a conspicuous group of interesting "organisms" that actually consist of the combination of a fungus and a photosynthetic organism, usually a green alga, growing in a symbiotic, or mutually beneficial, relationship. The fungal component of most lichens is a sac fungus, though a few are club fungi. In addition to the 30,000 known species of sac fungi, an additional 20,000 or so species of sac fungi occur as lichens. Lichens are familiar as gray-green old-man's beard hanging from trees in cool, moist climates and red-topped, soil inhabiting ones commonly called British soldiers. Lichens produce a great variety of chemical substances, some of which have been employed as natural dyes.

Although the club fungi and the sac fungi are characterized by the type of sexual spore they produce, the sexual phase of many fungi is unknown, and these are known as the Fungi Imperfecti, or imperfect fungi. Many fungi fall into this category; some 25,000 imperfect fungi are known. Because of their structural similarity to sac fungi, most are thought to be members of that group, while some are thought to be club fungi. Many imperfect fungi are useful in fermentation processes, commercial production of chemicals, and as producers of medicines. Penicillin, the first antibiotic, which was developed shortly before World War II, is the product of an imperfect fungus.

PLANTS

Having described and discussed several groups of organisms that are no longer classified as plants, we are left to define and categorize the kingdom Plantae. Even in the broad listing of kinds of plants given earlier (pines, palms, oaks, privets, roses, honeysuckles, marigolds, lilies, orchids, asparagus, oranges, apples, and pineapples, and those less familiar ones, cycads, ferns, horsetails, mosses, quillworts, liverworts, and hornworts), certain shared characteristics make them plants as well as differentiate them from the other organisms just discussed, all of which are not plants.

First, they are all eukaryotes, having organized nuclei. This differentiates them from the prokaryotes, bacteria and cyanobacteria, but it does not differentiate them from protists and fungi.

The plants listed all share these features: their cell walls are made of cellulose; their cells contain plastids, the most conspicuous and important of which are chlorophyll-containing chloroplasts; and they store the products of photosynthesis as starch. These features amply separate them from the fungi, which are, of course, also eukaryotic.

It remains then to differentiate the plant kingdom from the protists, a kingdom not as neatly defined as some of the others. There is great variation in the kinds of photosynthetic pigments found in the various groups of protists, while some contain no photosynthetic pigments at all. Cell wall composition and food reserves also vary. But the combination of these characteristics found in plants is not present in the protists, with the exception of some of the green algae, Chlorophyta. Thus while some protists store carbohydrates as starch, none store their starch in the chloroplast and none of the chlorophyll-containing protists contain both chlorophyll *a* and *b*, except some Chlorophyta.

So there are a number of similarities, but more important are the ways in which plants and protists differ. Plants have embryos; Chlorophyta do not. Embryos are young plants resulting from sexual union that remain attached to, surrounded by, and receive nourishment from specialized tissue of their mother. Unlike Chlorophyta, plants are fundamentally organisms that inhabit the land, and the presence of embryos is one of many adaptations that plants exhibit that adapt them to the harsh conditions found on land.

In the aquatic environment water provides support, reduces light intensity and quality, reduces fluctuations in temperature, and provides a medium through which nutrients flow. The medium of the terrestrial environment, air, provides none of these things. Here, environmental light is intensely bright and contains the full spectrum of wavelengths, including some that are harmful to living tissue; temperature varies enormously, sometimes over very short periods of time. Therefore, plants that live on the land have evolved features to compensate for the lack of water as an environmental factor; some of these features are readily evident in the plants we see daily. Accessory pigments may absorb harmful light. Sturdy stems support leaf surfaces, where photosynthesis usually occurs, and they contain specialized tissue that transports water and nutrients from the soil and the products of photosynthesis away from leaves.

Tracheophytes and bryophytes are plants. Tracheophyte means tube or vessel plant, referring to the usually complex internal system of specialized cells that transport water, nutrients, and other substances in these kinds of plants. The term *vascular plant* is more commonly used. In all vascular plants the conspicuous or dominant generation is the sporophyte or asexual plant. Bryophytes, or moss plants, are the mosses (about 10,000 species) and the similar liverworts (6,000 species) and hornworts (100 species). Most of these plants lack complex internal conducting tissue, and, more importantly and more interestingly, the gametophyte or sexual generation dominates.

Most accounts, popular or scientific, begin their discussions of mosses by saying that they are small plants that grow in moist, shady places. In fact, many are small and many do grow in moist, shady places, but some are large and many occur in usually dry, sunny places like deserts.

Mosses are leafy. The leafy plant of a moss produces gametes or sexual cells in specialized organs. These gametes, sperms and eggs, fuse to produce the sporophyte or asexual phase of the life cycle. The sporophyte of a moss usually consists of a foot, sunk into the gametophyte, a stalk or seta, and a capsule, in which the asexual spores are produced. Mosses typically produce thousands of spores in each capsule, and they are gradually released by the capsule rather than being released all at once.

Mosses do not produce a great variety of chemical substances nor do they store food or other organic products in quantity. Therefore they are not important sources of useful chemicals or food. But in surveys made over the past few years there are indications that some mosses contain substances that inhibit the growth of certain kinds of cancer tumors. Since mosses have not been tested for their efficacy against other diseases, there may well be useful substances to be found in this group of plants. Certainly the greatest use of mosses is in horticulture. Each year thousands of tons of peat mosses, *Sphagnum,* are harvested for use by professional and amateur gardeners.

Probably the most conspicuous role of mosses is as colonizers of new habitats, be these natural ones like newly formed volcanic islands or lava flows, or man-made ones like cracks in sidewalks or road banks. Mosses are often dominant among the first inhabitants of such habitats. Their presence provides pockets of moisture and humus in which other plants can become established.

The remaining plants are vascular plants, those with usually well-developed internal conducting tissues in the sporophyte, or asexual phase. *Conducting tissues* are groups of specialized cells that are involved with moving substances inside the plants. There are two kinds: the xylem, that conducts water and minerals, and the phloem, that conducts the products of photosynthesis. In vascular plants the sporophyte is dominant and free-living, unlike the bryophytes. The vast majority of the plants that we encounter day to day are vascular plants, which provide many of our needs. Vascular plants are the only plants that produce seeds, but not all of them do so. The ferns and similar plants are seedless, while gymnosperms, the conifers and similar plants, and angiosperms, the flowering plants, produce seeds.

With two minor exceptions, the vegetative bodies of all vascular plants are differentiated into three parts—roots, stems, and leaves—though because of specialization of function or appearance, they go by different names. For example, the leaves of ferns are often called fronds.

Roots serve to attach plants to their substrate, usually but not always soil, and to absorb minerals and water from it. Stems are usually above the substrate and provide support for the leaves and reproductive organs of the plant. The conducting tissues of the stems move minerals and water from the root system to the leaves and move the products of photosynthesis, which occurs mostly in the leaves, to other parts of the plant.

While there is tremendous superficial and microscopic variation in the roots, stems, and leaves of vascular plants, in their modes of reproduction lie the most fundamental differences between the various groups.

Ferns, with about 12,000 species, are the most abundant of the vascular plants that do not reproduce by means of seeds. The fern plant is the sporophyte and produces clusters of specialized groups of cells, called *sporangia.* Certain cells in the sporangia produce usually wind-borne spores. The spores germinate to produce small, free-living (photosynthesizing) gametophytic plants. These plants in turn produce gametes, eggs and sperms, that fuse to produce a zygote, initiating once again the sporophytic generation. The young sporophyte is initially attached to the parent gametophyte but soon overgrows it.

Several other groups of vascular plants reproduce in this fashion. These are often called fern allies, although they are not particularly closely related to the ferns. None of these groups contains many species. They are the horsetails or scouring rushes (*Equisetum*), with about 15 species; the 1,000 or so species of *Lycopodium* (club-mosses), *Isoëtes*

Moss, Pilotrichum hypnoides. *In* Descriptio et Adumbratio Microscopico-analytica Muscorum Frondosorum *by Johann Hedwig (1785–1797).*

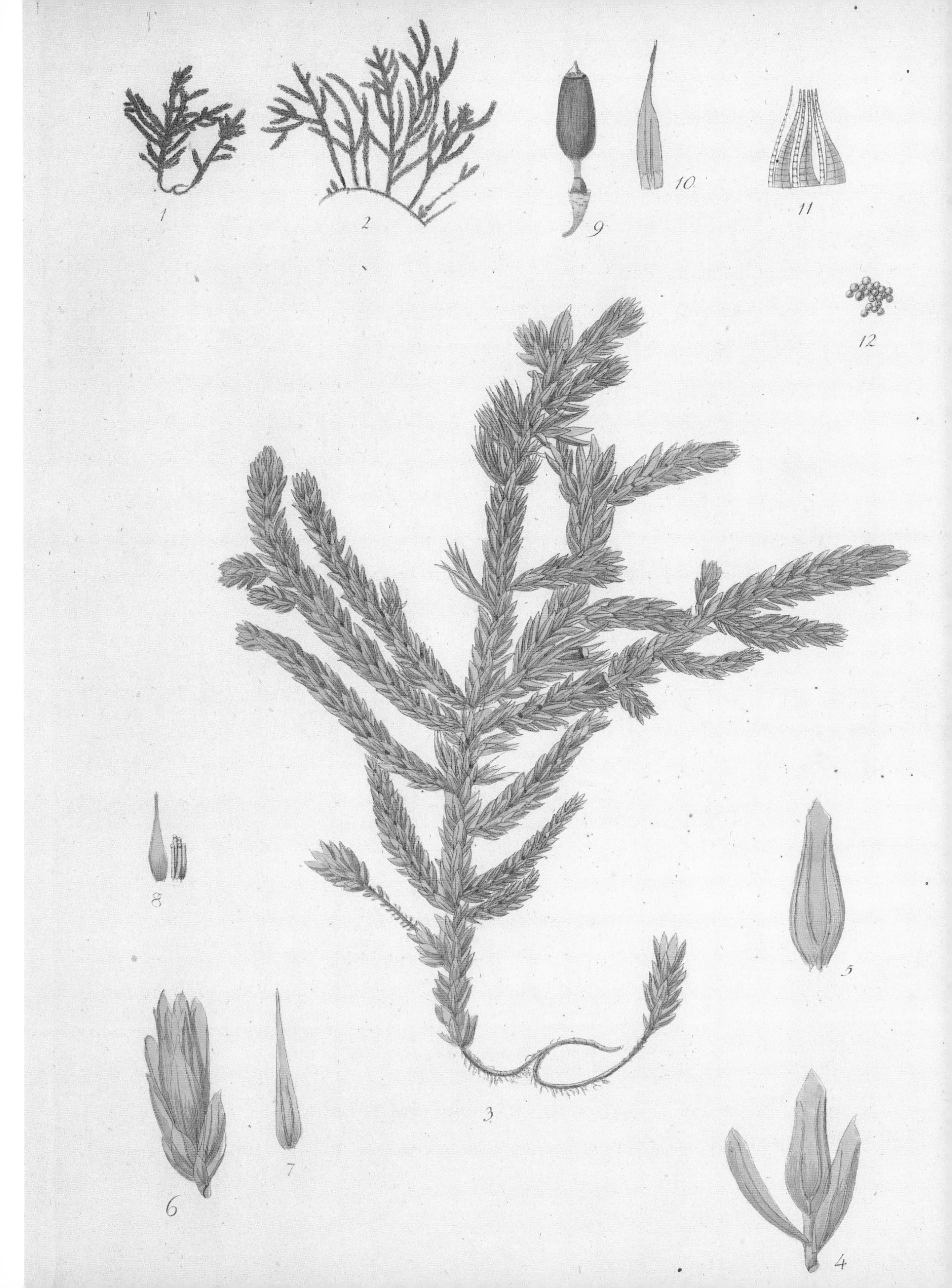
1
2
9
10
11
12
3
8
5
6
7
4

(quillwort), and *Selaginella;* and the 10 or so species of *Psilotum* (whisker fern) and *Tmesipteris.* The resurrection plant of the southwestern United States is a species of *Selaginella.*

The final group of plants are the seed plants, by far the most commonly encountered and abundant plants on Earth. As in the ferns, the asexual or sporophytic plant dominates, and the sexual plant or gametophyte is reduced, in this case to or nearly to microscopic size; it is neither free-living nor does it contain chlorophyll. The seed, which gives this group its name, is a complex feature, originating from both sporophyte and gametophyte. The seed contains the embryo, which upon germination begins to grow into a new plant. Quite often a supply of food is stored in the seed, which nourishes the young plant in its early growth phase.

There are two distinctive groups of seed plants—the gymnosperms, conifers and their relatives, and the angiosperms, or flowering plants. Gymnosperm simply means naked seed, and reflects the fact that the seeds of these plants are produced naked on the surfaces of specialized structures, often scales. Angiosperm means vessel seed, and this group, the flowering plants, produces its seeds enclosed in tissue, the ovary, which is part of the flower.

All gymnosperms are woody and are trees or shrubs in growth habit. Their leaves tend to be thick, and many are evergreen. There are less

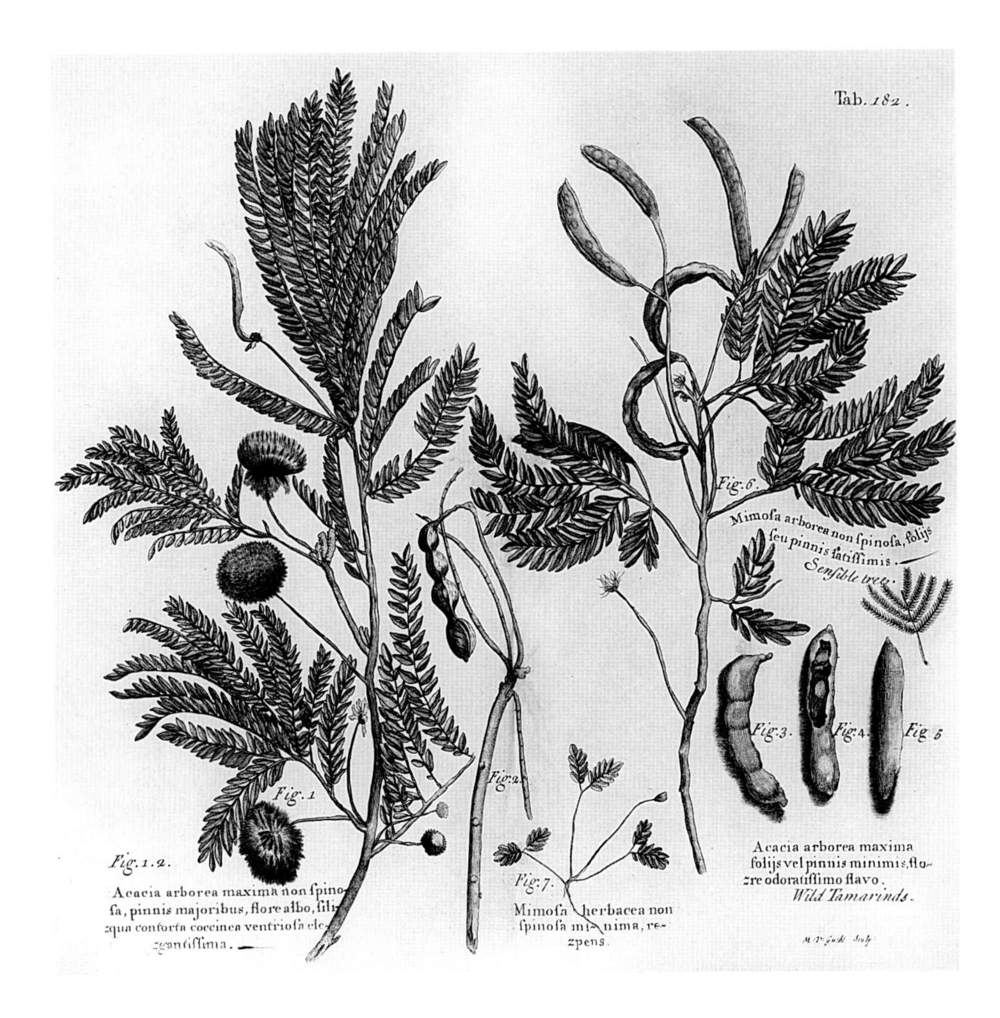

Legumes. In A Voyage to the Islands of Madera, Barbadoes, Nieves, St. Christophers and Jamaica *by Sir Hans Sloan (1725). The polynomial system of naming plants, which employed long descriptive phrases, is evident in this plate of several West Indian legumes.*

Opposite:
Amazon water lily, Victoria amazonica. *In* Victoria regia, or the great water lily of America *by John Fisk Allen (1854).*

than 1,000 gymnosperms on Earth. The largest group is the conifers, with about 550 species, most of which occur in north temperate latitudes. Conifers usually produce their seeds on the scales of cones, and pines provide a well-known example. Other groups of gymnosperms include the cycads, with about 100 species, mostly of tropical and subtropical areas. The maiden-hair tree or ginkgo (*Ginkgo* in botanical parlance) is also a gymnosperm.

Conifers tend to occur in cool climates. Large portions of the northern latitudes of North America and Eurasia are covered with forests in which conifers such as firs, larches, and spruces dominate. In warmer areas, conifers often occur at higher altitudes, where the local climate is cooler. They are nearly absent from the tropics. Conifers are important sources of wood, particularly lumber used in construction, and as sources of pulp for the manufacture of paper.

Angiosperms or flowering plants are the most abundant kind of plant now on Earth with something on the order of 235,000 species. Their characteristic feature is the flower, crucial to their reproduction through the production of seeds. Flowers vary enormously in size, shape, complexity, and color, and it is this feature that forms the basis for the overall classification of flowering plants. Flowers consist of sepals, often green, and petals, usually colored, both sterile organs, and stamens, the pollen-producing male parts, and carpels, which contain the ovules that mature into seeds.

The enormous diversity of flowering plants has been exploited by mankind, but only a small portion of it has been sampled in an organized way. For instance, most of our major crops were selected by primitive peoples, long before the development of the scientific method. Given the demonstrated benefits of using flowering plants to satisfy so many of our basic needs, surely a great treasure of new resources lies in this group of plants. Flowering plants are most diverse in the tropics of the world, the areas that are threatened with destruction in the near future.

Our collective ability to solve the problems of how best to deal with these threats to plant life, which provides for so many of our basic needs and which holds so much promise for helping solve many of the problems facing us, depends on continued support for the research being carried out at the Missouri Botanical Garden and at similar institutions around the world. Many thousands of plants will certainly be lost through extinction, but there is still time to detect many as yet unknown ones, some of which will survive and add to our store of useful plants. Other plants, poorly known now, will continue to be investigated and have uses found for them. The road to this knowledge of plants begins with the accumulation of fundamental information about them. This research, beginning with characterizations of plants, their naming, and their classification into related groups as genera and families, is not simply an academic exercise, carried out to help satisfy mankind's natural curiosity about the world around us. Rather it forms the basis for all further understanding of plants.

An oriental sweet gum, oak trees, and bottlebrush buckeye in full leaf.

HENRY SHAW AND THE EARLY FOUNDATIONS OF THE MISSOURI BOTANICAL GARDEN

In the midwestern city of St. Louis, Henry Shaw lived in a beautiful world: it was eighty acres big and filled with plants from all over North and South America, from Europe and the Mediterranean. During his lifetime, Shaw shared his idyllic universe with visitors and upon his death one hundred years ago in 1889, he bequeathed it to the public. The magnanimous gesture was typical of the fascinating man whose name is forever linked with the Missouri Botanical Garden.

Central to Shaw's hard-driving ambition was the frontier spirit by which he lived: no dream was too great, no hope impossible to implement. It was a sign of his aspirations and perspicacity that this Englishman, whose arrival in America at the age of eighteen to salvage his father's floundering cutlery business, accomplished that with remarkable skill and in good order, and something much more significant. In his adopted city of St. Louis, Shaw intended to duplicate the splendid English gardens at Chatsworth, the magnificent 1,100-acre estate southwest of Sheffield, the place of his childhood. That plan quite literally grew into a world-renowned botanical garden in its day, and now the oldest institution of its type in the United States. Moreover, Shaw also gave additional land to the City of St. Louis in 1868 for Tower Grove Park and endowed Washington University's School of Botany in 1885. His was the legacy of a successful entrepreneur turned philanthropist.

A key to Shaw's success was his obsessive personality. When he decided to see the world, he traveled for eleven years in the United States and in Europe. When he established a home, he built not one but two mansions: a city home on Seventh and Locust streets and a country home, which is still preserved and open to visitors on the grounds of the Missouri Botanical Garden. Shaw's style was to scrutinize all the aspects of a project—such as studying the best markets to barter tobacco and cotton for his business or researching ginkgo trees to enhance his garden—and then to act swiftly and decisively.

Opposite:
The proud owner-philanthropist, represented by a depiction of the Conservatory on the left and Museum Building on the right, in a watercolor by Emil Herzinger, 1860.

Overleaf:
By 1875, when this topographical survey was drawn by Camille Dry, Shaw had already established several buildings on his botanical paradise, among them his residence and mausoleum, the Palm House, Pavilion, Casino, and various lodgings for the caretakers of the Garden. This image is taken from Pictorial St. Louis, *edited by R. J. Compton.*

TOWER GROV
MAGNOLIA

Manhood for Shaw began in England with worries, not dreams. His father, Joseph, a once-prominent iron factory owner, had gone into debt. Desperate for funds, the elder Shaw borrowed money for a new venture which likewise failed, whereupon Joseph and Henry Shaw fled England to avoid creditors. Henry was upset over his father's "anxiety" to provide for his family. "Nothing would give greater satisfaction to me than to see my dear parents happy . . . in their declining years," the dutiful son wrote to his father on August 20, 1820, after leaving to seek his fortune in the American West. By that time, Shaw assumed the responsibility of supporting not only himself, but his father, mother, and two sisters. From these early experiences he gained a valuable lesson: he never borrowed funds.

Shaw, the first child born to Joseph and Sarah Hoole on July 24, 1800, attended Mill Hill secondary school, twenty miles north of London. There he received religious training and a classical education. He excelled in mathematics and French, which later helped him deal with traders on the American frontier. While other boys played cricket, Henry, like his father, preferred toiling in the garden, according to a history of Mill Hill.

Shaw's arrival in America began by way of Canada. It was to Quebec that he and his father set passage from England in 1818; although young Shaw had not graduated from school, it had become necessary to avoid his father's business creditors. While his father remained in Canada, Henry journeyed by sled across the trails of upper New York State, making a connection by boat to New Orleans. There he met up with a consignment of cutlery from Sheffield, with the hope of establishing an import-export business. But as he explained in a letter to his family on July 5, 1819, "the cutlery market was glutted" and "the humid, long Louisiana summers" were not to his liking. Pushing forward yet again, this time the young adventurer boarded the *Maid of Orleans* for St. Louis, having heard that it was an outfitting port for the frontier where business could be conducted in French. Spurred by the desire to reconcile his family's financial problems, Shaw settled in and got right to work.

Some time later in 1819 he established a wholesale business supply on the second floor of a building in downtown St. Louis. There he stocked two-thirds English goods and one-third American wares, selling ten to fifteen percent below the usual prices. Not only did he display his inventory of saws, knives, scissors, and other cutlery at this storefront, but he took his meals there as well. The first-year profits were small, and helping his father seemed a distant hope: ". . . every cent I can muster will be required in making a sufficient remittance for Sheffield. I have already been ashamed of . . . the very ill success I have met," he wrote.

Fortunately, his concerns were short-lived. He had opened his hardware store in the right place. Pioneers heading west purchased his hatchets and knives. Indians and fur traders bought his traps. By the late 1820s, he was regularly dispatching money to his family, who had by then settled in Rochester, New York. In a warm letter to his son on August 31, 1831, Joseph Shaw thanked him for the hundred dollars he had sent, "the numerous remittances" from some years past, and for supporting his mother and sisters "in so comfortable a manner, since I myself have been rendered incapable of providing for my family, or even of contributing in any measure. . . ." By the 1830s Shaw had diversified into exporting tobacco and cotton, trading furs and lead, selling supplies to the United States Army, and trading with the Indians. He wisely plowed his profits into real estate. At various times, he owned much of the property in South St. Louis and in the vicinity of the Garden, as well as considerable property in downtown St. Louis.

Shaw had a wholesome respect for money, bordering on penny pinching. Possibly even to save a few cents on postage and paper, Shaw would write the last page of a letter across the script of the first page. But his prudent ways in business paid off. Barely twenty years after his flight from England, he showed a net profit of $25,000. Though he continued to invest in other ventures, Shaw had earned enough money to allow him an early retirement at the age of forty. Now he had time for new challenges and adventures. The necessities of business sufficiently quelled, Shaw lavished money and time on travel, new homes, and horticulture. Moreover, now the gray-haired businessman was at liberty to court women.

Romance, it seemed, was not a gentle art easily mastered. The most sensational of Shaw's romances was that between him and the attractive Effie Carstang. What had begun in private resulted several years later in a public showdown in court when, in 1858, Carstang sued Shaw for breach of promise. She charged that the two had agreed to marry, but that Shaw's ardor cooled in direct proportion to his growing passion for the Garden. Effie Carstang was awarded a hundred thousand dollars in damages, but following the court's agreement to a retrial, the second verdict was in favor of Shaw.

Although a bachelor, Shaw commissioned leading St. Louis architect George Ingham Barnett to design city and country homes. Shortly after his country villa was completed in 1849, Shaw made one of his frequent trips to London. While there in 1851, he visited the Crystal Palace Exhibition and the Royal Botanic Gardens at Kew. But according to his own statement, his first idea for establishing a botanical garden in St. Louis occurred to him when he was a guest at Chatsworth, the Duke of Devonshire's country seat in Derbyshire, one of the most sumptuous private residences belonging to a non-ruling family in Europe. Shaw identified with the gracious English setting in bloom around him, which is said to have reminded him of the enjoyment he felt as a child when he had gardened with his sisters.

The dream to develop a botanical garden on the property surrounding his country villa took shape. Upon his return home, he began planting thousands of trees and shrubs on approximately eighty acres of his country estate, part of an 1,800-acre tract he owned a few miles from what was at the time the outskirts of the city. Early on,

Shaw planned the Garden not only for his personal pleasure but for the public and the advancement of botany, thereby laying the groundwork for the present-day three-prong thrust of display, research, and education. Eventually his comfortable St. Louis estate began to take on the atmosphere of the magnificent English gardens he had seen. Typically throwing himself into the task, Shaw began to lay out miles of roadways and sturdy walkways—some, five layers thick. A massive stone wall was placed on three sides of the Garden to shelter the plant life from foxes, deer, rabbits, and wind. Today that same wall gives a feeling of seclusion from the hectic world outside.

Another immigrant to St. Louis, the German botanist Dr. George Engelmann, and Sir William Hooker, director of the Royal Botanic Gardens at Kew, encouraged Shaw to expand his original concept and to include greenhouses, a botanical library, a herbarium, and a museum in his plans. Shaw began working on these goals in 1851, completing them nearly a decade later. At the northeast end of the Garden, he erected arched trellises giving access to the walled fruticetum, a collection of shrubs and fruit trees. To the south of the fruticetum, the formal gardens were developed, including a greenhouse complex and a viewing "Pavilion." On the west side of his estate, the arboretum was built. This site contained most of the trees that grew in the Missouri climate, such as hollies, wild cherries, and sugar maples plus varieties from other states and countries that Shaw had seen in his travels, including cedars of Lebanon and maidenhair trees from the Orient. In 1857 Engelmann had purchased the herbarium of Johann Bernhardi, a German scientist, on behalf of Shaw. This required the establishment of a museum to house these 60,000 plant specimens and a library. It was completed in 1859, at a cost of $25,000, in time for the public opening of the Garden. Later, in 1882, the Linnean House, named for the Swedish botanist Carl Linnaeus, was constructed on the northeast section of the Garden as the display area for camellias. These early buildings and the elaborate entrance gates to the Garden were designed by Barnett. (Only the museum and Linnean House exist today. The other structures have been replaced by newer facilities including the Ridgway Center, the main entrance, and the Climatron, a geodesic-dome greenhouse, devoted to vegetation that thrives in tropical regions.)

By 1859 the Missouri Botanical Garden, affectionately called Shaw's Garden even today, was well underway. Shaw was ready to open it to the public six days a week and two Sundays a year, although expansion was still taking place. Just as Shaw had dreamed, the Garden not only provided a botanical experience for the public but a retreat from city life. Visitors were welcomed at the entrance along Floral Avenue, now Flora Place, with young plantings including poplars, weeping willows, Japanese anemones, and Norway spruces. People came to admire the formal rose garden and the circular beds of annuals, and to exclaim over the palms and other exotica. Both the well-to-do and the working classes mingled along the pathways beneath some of Shaw's favorite

A portrait made in 1860 of the great scientist, George Engelmann, by R. A. Clifford. Today the painting hangs in the Garden's Lehmann Building.

An early view of the Pinetum at the Shaw Aboretum, known for its collection of conifers.

Although Juno is now situated on the grounds near Tower Grove House, in earlier days she presided over the formal Italian Garden.

A stroll in Shaw's Garden was a pleasant way to spend a leisurely day. Here visitors make their way around the young plantings near the Conservatory, 1868. Juno is visible in the rear, center.

Neither trompe l'oeil nor carnival maquette, this brave little girl and the couple are actually poised (tentatively, no doubt) atop water lilies, around the turn of the century. The broad bases of the lilies were usually braced with boards so as to provide somewhat stable platforms for adventurous visitors.

trees—silver birch, white pines, elms, chestnuts, magnolias, and English yews. They arrived on foot, by streetcar, and by carriage to breathe the fresh air, to stroll among the floral paradise, and to have their pictures taken in their Easter finery or in their Sunday best. They still come today—over a half million visitors a year.

Despite Shaw's efforts in the first years, the Garden could not be called a full-fledged botanical institution, but it was a firm beginning. Just as he had done earlier in his business, Shaw made the commitment in time, money, and energy to insure its expansion. He educated himself on all aspects of his botanical paradise and had superior knowledge of trees, recognizing even rare specimens. He sought advice from the prominent men in the field, including Engelmann, Hooker, and the landscaper John Thorburn. In 1860, at the urging of Engelmann and Asa Gray, the famous Harvard botanist, Shaw hired Augustus Fendler, a plant collector, to catalogue the botanical specimens in the herbarium. (Fendler stayed for only one year.) While botanists then may have questioned Shaw's amateurish efforts and looked down upon him because he lacked academic credentials, *The Magazine of Horticulture* praised the Garden in its September, 1859 issue as "one of the most magnificent projects in this or any other country, to wit: the establishment and endowment, by private individual munificence, of a Public Garden on a broad and liberal plan . . ."

The Garden was recognized officially when the Missouri Legislature passed an enabling act on March 14, 1859, approving Shaw's intention to convey to the trustees 760 acres of land "situated in Prairie des Noyers Common Fields for the development and support of a botanical garden." To celebrate the fiftieth anniversary of his arrival in St. Louis, Shaw invited friends to the Garden on May 4, 1869. "There is not a person in St. Louis, we venture to say, who is not familiar with [the Missouri Botanical Garden] in Tower Grove," reported a story about the anniversary party in the *Missouri Republican.* It added that the Garden's principal charms were the conservatory and the beauty of the arrangements.

With the goals for the Garden in sight, Shaw turned his attentions to public parks in St. Louis. The city's taxpayers were slow to accept responsibility for large municipal parks. Thomas O'Reilly, a physician, suggested that Shaw make "a gift to the city of grounds for a public park." In 1868 Shaw did just that, offering the city a tract of nearly 280 acres adjacent to the Garden, running between Magnolia Avenue and Arsenal Street from Grand Boulevard on the east to Kingshighway on the west. In return, the city was to provide $360,000 toward the development of what would be known as Tower Grove Park. State law did not permit the city to accept the property since most of the land was then outside the city limits. Shaw solved the problem by persuading the state legislature to pass a law establishing a special board of commissioners to be appointed by the state supreme court, including himself and, following his death, the future directors of the Missouri Botanical Garden.

Opposite, above:
An illustration of Shaw's Garden as it appeared in The Illustrated Journal of Agriculture, *c. 1879–82.*

Opposite, below:
Visitors arrive by horse-drawn carriages at the Main Gate—the formal entrance to the Garden—in 1890.

ICE CREAM SALOON
RESTAURANT
TO TOWER GROVE PARK
NOT OPEN

Summer has always been terribly hot in St. Louis, and these ladies show good sense to open their parasols against the sun. The Observatory likewise offered refuge and a splendid view of the Garden.

Overleaf:
From this group of "loving hands" came the maintenance and upkeep of the grounds: garden workers gathered in front of the entrance to the Mausoleum grounds, 1890.

Barely visible in the foreground is Shaw, posed in front of Tower Grove House. At top is the encircled area where he would take measure of doings in the Garden below.

Contemporary views of Tower Grove House. Shaw's residence has been faithfully restored and maintained as an authentic display of the period.

Shaw in his middle years, as published in the Encyclopedia of the History of St. Louis, *edited by Hyde and Conard, 1899.*

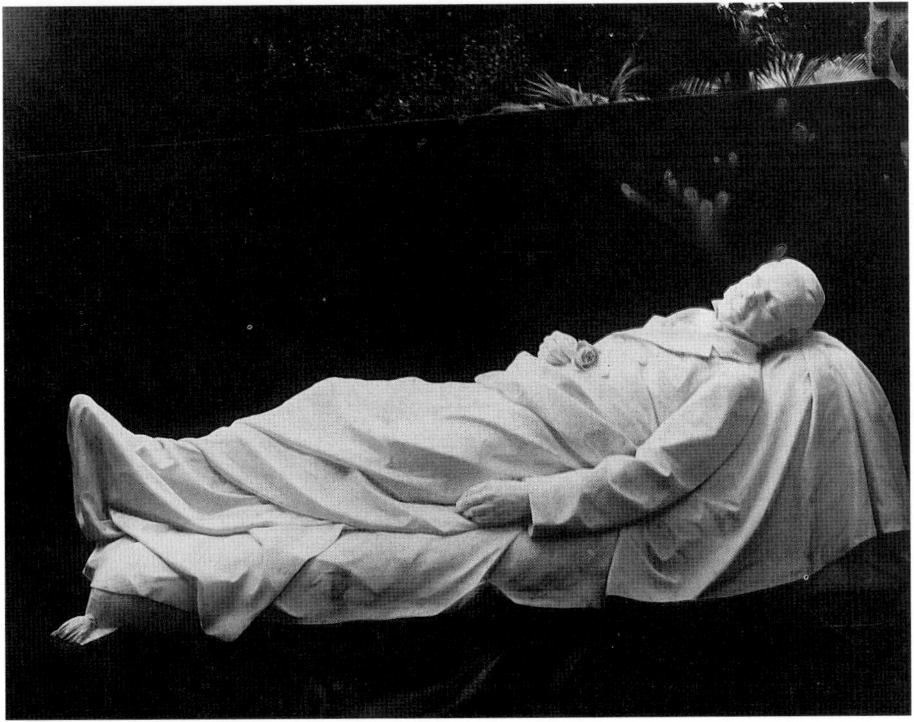

Shaw as he wished to be remembered for perpetuity: at rest in his mausoleum.

Opposite:
Dr. William Trelease was successor to Shaw as director of the Garden. This photograph, showing him at work in his office in Tower Grove House, was taken just three months after Shaw's death.

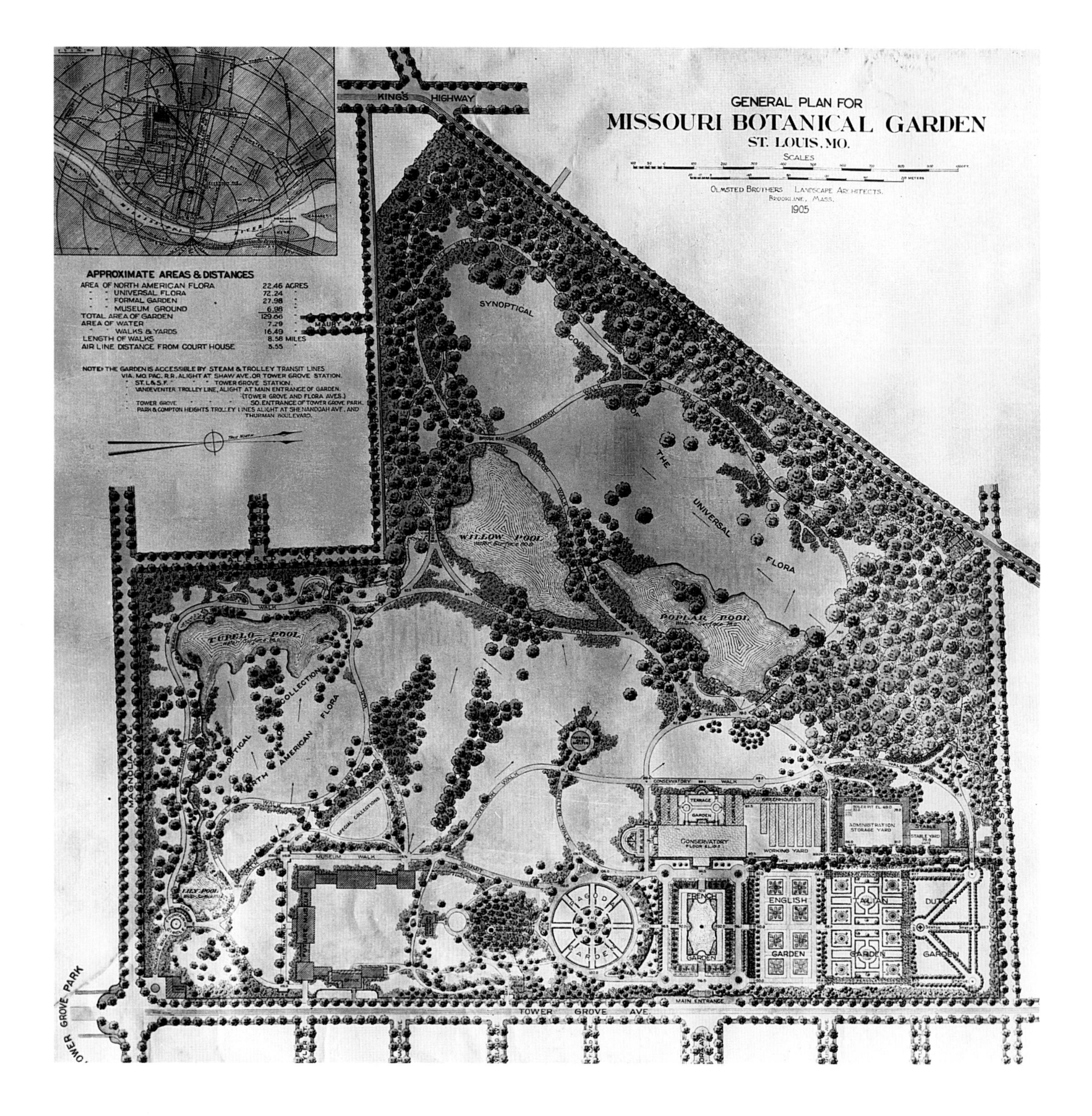

GENERAL PLAN FOR
MISSOURI BOTANICAL GARDEN
ST. LOUIS, MO.
SCALES
OLMSTED BROTHERS LANDSCAPE ARCHITECTS.
BROOKLINE, MASS.
1905
APPROXIMATE AREAS & DISTANCES
AREA OF NORTH AMERICAN FLORA 22.46 ACRES
" " UNIVERSAL FLORA 72.24 "
" " FORMAL GARDEN 27.98 "
" " MUSEUM GROUND 6.98 "
TOTAL AREA OF GARDEN 129.66 "
AREA OF WATER 7.29 "
" WALKS & YARDS 16.49 "
LENGTH OF WALKS 8.58 MILES
AIR LINE DISTANCE FROM COURT HOUSE 3.55 "
NOTE: THE GARDEN IS ACCESSIBLE BY STEAM & TROLLEY TRANSIT LINES
VIA. MO. PAC. R.R. ALIGHT AT SHAW AVE. OR TOWER GROVE STATION.
" ST. L.& S.F. " " " TOWER GROVE STATION.
" VANDEVENTER TROLLEY LINE, ALIGHT AT MAIN ENTRANCE OF GARDEN.
(TOWER GROVE AND FLORA AVES.)
" TOWER GROVE " " " SO. ENTRANCE OF TOWER GROVE PARK.
" PARK & COMPTON HEIGHTS TROLLEY LINES ALIGHT AT SHENANDOAH AVE. AND THURMAN BOULEVARD.
KINGS HIGHWAY
MAURY AVE
SYNOPTICAL COLLECTION OF THE UNIVERSAL FLORA
WILLOW POOL
POPLAR POOL
TUPELO POOL
SYNOPTICAL COLLECTION NORTH AMERICAN FLORA
SPECIAL COLLECTIONS
LILY POOL
MUSEUM WALK
CONSERVATORY WALK
TERRACE GARDEN
GREENHOUSES
CONSERVATORY
WORKING YARD
ADMINISTRATION STORAGE YARD
STABLE YARD
FRENCH GARDEN
ENGLISH GARDEN
ITALIAN GARDEN
DUTCH GARDEN
TOWER GROVE AVE.
MAIN ENTRANCE
MAGNOLIA AVE
TOWER GROVE PARK

Opposite, a plan made of the Garden in 1905 that was never executed, and, above, a realistic view of the Garden as it existed in 1913.

With his understanding of landscape planning, Shaw scattered pavilions, summerhouses, and statues of renowned men throughout the park. The blending of manmade sculptures and the cultivated landscape enhanced the park's beauty while evoking the grandeur of an historic past. What these eminent figures meant to Shaw can only be speculated upon. Included were Columbus, who like Shaw was an adventurer; von Humboldt, the famous naturalist; Shakespeare, one of Shaw's favorite writers; and several of Shaw's favorite composers. The ever-thrifty Shaw noted that luckily the art works were imported "duty free" since their use was for public pleasure. Tower Grove Park remains today, still governed by its special board of commissioners, as one of the finest Victorian parks in America, with over one hundred varieties of trees.

Shaw's regard for beauty and enjoyment was no less evident in his homes, especially his Italianate-style country villa, Tower Grove House. The residence, built in 1849 and situated near a grove of sassafras trees, had more than a few distinctive features, suitably odd for its principal resident. Overall its three-storied facade bore an asymmetrical design. For example, the servant's quarters in the eastern part of the house did not match that part of the house facing west. Shaw had an oblong window set over the front door, allowing for more light, perhaps, but also for a prominent view of the Garden. And atop the house's tall, square tower a rail surrounded the roof, where Shaw would observe, spyglass in hand, his garden contingent at work. Inside, the original home probably had two bedrooms, a kitchen, a study, and a library. Later, the trustees of the Garden enlarged the residence and added indoor plumbing (it had been inadequate in Shaw's day) to accommodate the first director of the Garden after Shaw's death, Dr. William Trelease, and his family. Although the eastern portion was rebuilt, the western section of the house is still the original structure. For his more traditional city house, cost and design were odd bedfellows. For example, Shaw purchased 300,000 bricks at $5.00 per thousand and bought mahogany lumber for 19 cents a board foot. While on his frequent trips to Europe, he purchased marble in Italy for the fireplaces. As a result of one of the bizarre stipulations in Shaw's will, the city house was moved brick by brick to the Garden after his death and today forms the northern portion of the Administration Building.

For the next forty years, Shaw spent his summers in the tastefully furnished Tower Grove House. Although Shaw had been close to his family, especially to his sister Caroline, he had few intimates. His interests centered more on the development of the Garden. Shaw lived until the age of 89, when on August 25, 1889, he died of malarial fever.

In his will Shaw made numerous minor bequests, with the greater part of his estate, valued at over two million dollars, left as an endowment for the Missouri Botanical Garden. True to his methodical personality, the future of the Garden was well planned. To insure high standards, it called for the Garden's chairman to be chosen by the seven trustees serving life terms and by three distinguished ex officio members of the community: the Episcopal bishop of Missouri, the chancellor of Washington University, and the mayor of the City of St. Louis. Later the president of the St. Louis School Board of Education was added. (Since then, slight changes have occurred in the governing structure, including the recent addition of a larger body of trustees, serving shorter terms.)

Near the end of his life in 1885 and upon the advice of eminent scientists, Shaw gave money to create the Engelmann Professorship of Botany at Washington University. Dr. Trelease was named as the first professor. The connection was not without consequences. After Shaw's death, Trelease became the first director of the Garden and is credited with making it a true scientific center.

In his will, Shaw provided instructions for his burial, which he had well anticipated. Several years earlier, Shaw had posed for a photograph lying in repose, his eyes closed and his hands clutching a rose; from this a full-length recumbent statue was sculpted. At his request the sculpture was placed on his granite sarcophagus, inscribed with his name, at a site just north of the Tower Grove House. By virtue of his will, he was to be the only person buried in the Garden.

No doubt Shaw would have been pleased with the outcome of his grandest dream—the Missouri Botanical Garden, which is still thriving and growing.

A TOUR THROUGH THE GARDEN TODAY

The City of St. Louis is about twelve miles below the confluence of the Missouri and the Mississippi rivers. In 1763, Pierre Laclède, then thirty-four years old, and his stepson, Auguste Chouteau, fourteen years old, chose that spot for a fur-trading post under the sponsorship of a New Orleans firm that held the monopoly for the entire Missouri region.

Soft rolling sheets of grass dotted with brilliant wild flowers stretched out before the first visitors. The French called these grasslands a *prairie,* meaning a European-style meadow. A survey map of 1767 labeled the upland behind St. Louis north to the Missouri River as *Prairie immense où on mettra une multitude d'habitants* (huge prairie that will hold many settlers).

Although a prairie was primarily grassland, it did not have to be treeless. On many hillside summits red cedars, *Juniperus virginiana,* stood in rows while post oaks, *Quercus stellata,* and black walnuts, *Juglans nigra,* dotted the scene below. The fifteen-square-mile area that today includes the Missouri Botanical Garden was called *Prairie des Noyers,* or walnut prairie, because of the early presence of these trees. A prairie was not a sea of grass that stretched to limitless horizons but was a distinct tract of land, marked by forests and identifiable boundaries. When Lewis and Clark left Missouri in 1804, they chose the term *great plains* for those vast seas of grass that were found further west, but eventually the word *prairie* came into common use.

When Henry Shaw arrived in St. Louis in 1819, he was eighteen years old. When he was seventy-nine he recalled the spring of 1820 and his first sight of the land that eventually became his garden: "From the village of St. Louis I came through the bushes, by a narrow path winding among the sink holes or natural depressions of the commons, to the elevated ground now called Grand Avenue, where, open to the river, a beautiful prairie extended westward, uncultivated, without trees or fences, but covered with luxuriant grass, undulated by the gentle breezes of spring, not a tuft of which can now be found."

Over the following seventy years he amassed a fortune and created a garden which would eventually rank among the greatest botanical

A border of hostas, Hosta × hybrida *'Cathayana', are in full bloom under a crab apple,* Malus *sp., in the Bulb Garden.*

resources in the world. He directed the development of the garden until his death in 1889, leaving a will that provided for the establishment of a garden "with the view of having for the use of the public a Botanical garden easily accessible, which should be forever kept up and maintained for the cultivation and propagation of plants, flowers, fruit and forest trees, and other productions of the vegetable kingdom. . . ."

The Garden opened to the public on June 15, 1859. Practically from the start it was a success; by 1868 over 40,000 visitors had attended the Garden. An engraving in *Compton's Pictorial St. Louis* (1876) pictures a tropical greenhouse, a series of parterre-style ornamental beds, a pagoda, and the elliptic shape of the mausoleum grounds where Shaw was later buried.

In the fall of 1907, the gardens were such a success that the annual chrysanthemum show exhibited four thousand mature plants consisting of some four hundred cultivars, all under a canvas tent illuminated by electric lights. The total attendance for that November alone was 34,439. There could be no question that the Garden had become a community phenomenon.

Like the trees and plants it emulated, the Garden grew. The palm house was built in 1912, and the nearby Italianate Gardens were regarded then as the finest formal parterre in the country. In the area that now includes the Lehmann Rose Garden, a two-acre ellipse called the Economic Garden was developed in 1914; it featured farm crops, herbs, and four model backyard gardens for the instruction of the visitors. But by 1925, city smoke was becoming such a problem that even the brass fixtures in the Garden's herbarium would tarnish overnight. The search for a tract of land to serve as a growing area led to the original purchase of 1,300 acres at Gray Summit, thirty-five miles southwest of St. Louis. And by 1928 the number of visitors to the Garden hit a yearly high of 468,072.

During the 1930s and 1940s the Garden was calm with few changes taking place. Then in the 1950s the Economic Garden was eliminated and half of the Italian Garden became part of the floor of the Climatron, which was completed in 1960.

The Garden was designated a National Historic Landmark in 1971. More development followed in the decade after 1972 with the creation of the Japanese Garden, the English Woodland Garden, and the Lehmann Rose Garden.

Today with a roster of more than twenty thousand people, the Missouri Botanical Garden has the second largest membership of any botanical garden in the world.

THE GARDEN

With its international reputation, the Garden attracts people from all over the world; naturally some are there to enjoy the quietude that the enclosed grounds provide, while others are there to learn about the

Ridgway Center as a beacon of tiny lights against the St. Louis sky.

many kinds of cultivated plants in the vast display gardens. But no matter what purpose the visitor may have—as tourist or as professional—the first sign of welcome to all is the Ridgway Center.

Completed in 1982, this light-filled modern shell with its great vaulted ceiling reminiscent of a zeppelin serves a number of functions. It houses the Garden's floral display hall used for seasonal exhibitions. The educational facilities are here as well as an auditorium and some staff offices. The Garden has a long-standing emphasis on research, display, and education, and Ridgway Center is perhaps the building that best signifies this to the visitor.

Beyond the Center is Spoehrer Plaza, a paved open space that surrounds a fountain, ringed with Kentucky coffee trees, *Gymnocladus dioicus,* and flower beds that are changed seasonally.

AZALEA-RHODODENDRON GARDEN

West of the plaza is the pathway through the Azalea-Rhododendron Garden. These colorful shrubs are both members of the same plant genus, *Rhododendron.* Garden azaleas are largely deciduous, losing their leaves during the winter months, and their flowers are usually funnelform with a floral tube that gradually widens like a morning glory; rhododendrons are mostly evergreen and have bell-shaped flowers. The Azalea-Rhododendron Garden includes about one hundred cultivars; included in the display are an azalea bowl, a rhododendron glen, and a glade of magnolias. The size of the garden is a little over an acre and has been in place for about four years.

Although a few plants will throw out a flower now and then, azaleas and rhododendrons are really spring flowers. Spring in St. Louis starts in late March and runs through April and, depending on the weather, sometimes into May. When in bloom, the Azalea-Rhododendron Garden is awash with tints of ivory and apricot, pastel pinks, and deep purples.

Presently there are ferns including Christmas fern, *Polystichum acrostichoides,* maidenhair, *Adiantum pedatum,* and the lovely Goldie's woodfern, *Dryopteris goldiana.*

THE CLIMATRON

The crystal-like dome called the Climatron resembles the futuristic cities of science fiction. Viewed from a distance it rises like an enclosed weather-free World of Tomorrow, a symbol of the best of America's future. In its singular style the dome of this greenhouse has become a symbol for the Missouri Botanical Garden and the City of St. Louis.

With the approach of the Garden's centenary in 1959, the Trustees decided to replace the old Palm House with a new conservatory. The greenhouse they envisioned had to be up-to-date employing the newest

The Azalea-Rhododendron garden in May. The tree is a magnolia, Magnolia × loebneri, *and Korean rhododendrons,* Rhododendron mucronulatum, *bloom in the foreground.*

One of the first trees to bloom in the Garden is the magnolia, Magnolia *'Randy'.*

The futuristic dome of the Climatron is partially reflected in one of the Garden's several water lily pools.

principles of architecture. Reconciling desire and budgetary means, the design that eventually emerged was based on the ideas of R. Buckminster Fuller and his geodesic dome, having no internal columns or other supporting structures to stand in the way of the plant displays. The broad open space protected by its delicate framing suggested a step into the future.

The architectural firm of Murphy and Mackey was hired as a consultant. There were two problems to solve: how to rest the dome on the ground and what material should be used for the enclosure's surface. They chose aluminum tubing for the frame connected with a "skin" of lightweight clear plastic to allow the maximum penetration of sunlight. Because each hexagon in the dome's design turned out to be a slightly different size, a computer was used to chart the dimensions.

The superstructure consists of tubing arranged in hexagonal patterns, lined with a layer of transparent ¼-inch-thick Plexiglas, which was suspended just below the underside of the dome framework. Originally the idea was to hang a clear plastic sheet from the tubing, but no manufacturer would guarantee such a skin to hold up for more than a year or two.

The result is a dome-shaped greenhouse that rises 70 feet in the center and is 175 feet in diameter, enclosing a volume of 1,300,000 cubic feet and covering a ground area of 32,350 square feet, over three-quarters of an acre of land. The weight of 730,750 pounds of aluminum is carried to the ground along five main arching supports or "lunes" that radiate from the pentagon at the dome's center. There are no interior supports. Ultimately there were 3,625 individual panes of Plexiglas in 21 different sizes and 7 basic shapes.

Plants breathe. Their leaves release oxygen, carbon dioxide, and moisture, and inside an enclosed dome there is a tremendous volume of air that must be moved about and temperatures to control. To handle this, there were elaborate air circulation systems, with both temperature and air volume controlled by a complex of pneumatic switches and motors that operate a system of fans and dampers.

On October 1, 1960, the Climatron (the name was coined by Frits Went, the Garden's director during the 1960s) was dedicated to American Science, and to the people of America and St. Louis. With over 2,500 people in attendance, 100,000 watts of lights were turned on and the Climatron, rising like the eye of a gigantic insect, sparkled against the night sky. The following day, a Sunday, the Climatron opened to the public exhibiting the Garden's new collection of palm trees (those from the former Palm House were lost to snow and cold weather while the new dome was being built), tropical and semi-tropical plants, and many of its world-famous orchids.

The amazing Climatron became a public relations coup for the Garden. In many cases, people came just to see the dome which had become a prototype for things to come; the very name brought visions of mankind's future to the mind's eye. The yearly attendance at the Garden was 280,000 people in 1959. In the inauguration year of the Climatron, the figure jumped to 425,000 visitors.

A workman inserts a triangular Plexiglas panel in the aluminum framework of the original Climatron in 1959. The ¼-inch-thick Plexiglas was held in place with neoprene gaskets.

For their efforts, the architects Eugene J. Mackey and Joseph D. Murphy were the first Americans to win the R. S. Reynolds Memorial Award for the most significant work of architecture in which aluminum was a contributing factor. In the bicentennial year of 1976, the American Institute of Architects named the Climatron as "one of the most significant architectural achievements of the first 200 years of American History."

Over the years the Climatron began to dim and lose its sparkle. The action of the wind and weather in combination with the heat of the sun began to tarnish the Plexiglas. The gleaming surface cracked and dulled and by the end of 1987 only fifty percent of the daylight could reach the plants below. Rain seeped into the dome because of leaking joints. The dome was aging and needed care.

Beginning in the spring of 1988, the worn panes of Plexiglas were removed, and by late October were replaced with "Low E" glass, a new product intended to conserve solar energy. Like automobile windshields, the new glass will not fracture into deadly shards when broken, but will break into large dull-edged granules that will not cut. Each piece is formed from two individual panes of glass with a thermoplastic layer sandwiched in between.

Because the dome is a historical landmark the outside tubing of the old Climatron has been left in place, and a whole new support system for the glazing has been put up inside. The shapes of the panels are triangular and trapezoidal; six of each fit together to make up each hexagon. Internal gutters have been added to carry off condensation.

About sixty trees were protected but left in place during the construction as they were too large to move. Some are over two hundred years old and were displayed at the St. Louis World's Fair of

The forest floor of the Climatron.

1904 (more properly known as the Louisiana Purchase Exposition). Over 1,400 species of plants are at home in the new Climatron.

The remodeled interior of the Climatron focuses attention on three principal aspects of tropical rain forests: the understory (forest floor), the canopy, and the epiphytes (plants that grow on other plants, such as orchids). There will be a representational volcanic ridge, made of fiberglass and reinforced concrete, covered with vegetation indigenous to the tropics. Included will be tropical plants important to the economy of the world: bananas, cacao, coffee, and rubber, plus a selection of the Garden's important collection of orchids. Visitors will walk under waterfalls and around pools and a bog, amble along a catwalk through the tree canopy, and descend the mountainside to the world below.

From a concealed place in the basement of the dome, modern, computerized controls operate a new climate system, which includes the "cannon fan," an apparatus that blasts air from an opening in the mountaintop up to the roof. Noise is masked by the sound of three waterfalls cascading down the mountainside.

It is interesting to note that the new Climatron uses the Integrated Pest Management (IPM) system to control the problem of insect pests that continually threaten an enclosed tropical atmosphere. Instead of using chemical insecticides to fight spider mites and mealybugs, the Garden employs an army of predaceous mites, mealybug destroyers, and a division of ladybugs to wage the fight.

Opposite:
Within the Climatron a screw pine, Pandanus copelandii, *exhibits its stilt roots. Since these plants often grow in unstable soils on sea shores or in the mountains, such roots enable them to stand erect.*

Shielded from the fluctuations of temperature in St. Louis, many Climatron orchids bloom throughout the winter. The blossoms are a horticultural hybrid of the orchid genus Cattleya.

The male cones of the cycad, Encephalartos hildebrandtii.

The Brookings Interpretive Center

Upon leaving the Climatron, visitors pass through a new facility, the Brookings Interpretive Center. A large three-dimensional diorama gives them a firsthand glimpse of the devastation involved with the destruction of a tropical rain forest. Accompanied with the sound effects of a bulldozer's roar, they hear the crash of a giant tree being uprooted. Exhibits explain the results of this careless handling of the world's environment—at the current rate these rain forests will completely disappear within sixty years—and the disastrous impact on the world's climate. In fact, every minute an area of tropical rain forest about the size of the Climatron is destroyed.

THE TEMPERATE HOUSE

Until recently, the building in this area, opened in 1913, was known as the Mediterranean House and flanked the Climatron to the north. It has been replaced by the present Temperate House. The plants and exhibits in the Temperate House are covered by a shape designed to suggest the curve of the Climatron. The displays focus on plants from warm temperate regions. In addition to plants from the Mediterranean basin, are plants from Africa, Australia, South America, China, Korea, Japan, coastal California, and the southeastern United States.

Indoors a hillside is planted with wild flowers; an area devoted to plants of the Bible; a grape arbor; and a formal Persian garden with a fountain, all inside a walled garden surrounded by Italian cypress trees. As in the Climatron, rocks and rills are made of fiberglass-reinforced concrete; visitors can climb steps cut directly into a rock cliff. Instructive displays attest to the continuing threats to the environment of the temperate regions of the world.

THE ROCK GARDEN AND DWARF CONIFER GARDEN

Alan Godlewski, the Garden's late chief horticulturist, who died in 1988, designed the present Rock Garden and the accompanying Dwarf Conifer Garden, located in front of the Temperate House.

The Rock Garden was first began in 1972 but formally became an organized garden in 1982. It lies on either side of the Conifer Garden. Measuring about one thousand square feet, it contains a bewildering variety of plants that cheerfully bloom within, without, and around a field of flat-topped rocks and small and round boulders; even in October it is in full bloom. Amazingly resistant to drought, alpine plants and various wild flowers from around the world are quite at home in rock gardens, requiring little care except that the spare earth they are planted in provide perfect drainage.

Opposite:
Tropical trees against the brilliant light that falls through the skeletal framework of the newly renovated Climatron.

Overleaf:
Summer in the Rock Garden. An ornamental grass, Miscanthus sinensis *'Gracillimus', hovers above a clump of an annual cosmos,* Cosmos *'Sunny Yellow' and three patches of sweet alyssum,* Lobularia maritima.

Above:
Flowers of the Engelmann daisy, Engelmannia pinnatifida, *a genus of plants from America's Midwest, blooms in the Rock Garden throughout the summer.*

Opposite:
The white evening primrose, Oenothera speciosa, *is a native wild flower, which opens its pale-pink petals in the prairies and along the roadsides of the Midwest.*

Overleaf:
There are many unusual plantings in the Rock Garden. Here the tiny white flowers of sweet alyssum, Lobularia maritima *'Oriental Night', bloom above Dahlberg daisies,* Dyssonia tenuiloba, *their flowers in stark contrast to an ornamental grass,* Festuca ovina *var.* glauca.

Pages 88–89:
In the Dwarf Conifer Garden, at left a dwarf Norway spruce, Picea abies *'Microsperma', with* Verbena *'Hanging Basket Purple' at the bottom, and at right, grown from seed,* Chrysanthemum × morifolium *'Golden Dream'.*

Although most rock gardens are at their best in early spring, the Garden's collection of plants provide bloom from April to October. Included are Alpine speedwell, *Veronica incana;* the red cranebill, *Geranium sanguineum;* the yellow-flowered *Achillea taygetea* 'Moonshine' from Greece; Engelmann's daisy, *Engelmannia pinnatifida,* an American native that forms 3-foot-high mounds of flowers; and the exquisite Missouri evening primrose, *Oenothera macrocarpa.*

A number of ornamental grasses wave their feathery plumes beginning in September and include feather grass, *Pennisetum alopecuroides,* and the graceful maiden grass, *Miscanthus sinensis* 'Gracillimus', their leaves bending over the sulfur-yellow blossoms of *Coreopsis* 'Moonbeam'.

After the perennial plants die back in early winter, the frost-tinged leaves are left for the winter, to be cleaned out in early spring unless they exhibit disease. The effect is of a dried bouquet with the grasses, sedums, and veronicas forming their own winter decoration.

The Dwarf Conifer Garden is planted in a triangular plot of land, about 160 square feet, surrounded by the Rock Garden. It was completed in 1984. Siberian iris, *Iris siberica;* pasque flower, *Anemone pulsatilla;* a seed-grown chrysanthemum, *Chrysanthemum* 'Golden Dream'; and a beautiful flowering onion with rose-tinted flowers, *Allium senescens* (*A. glaucum*), which blooms in the fall, are all dotted about a cluster of twenty-three conifers.

Unlike their larger relatives, the dwarf conifers are more in scale with today's smaller home garden, many of them only achieving a growth rate of one inch per year. With great age, our native white pine, *Pinus strobus,* can reach a towering 120 feet, but the Garden's dwarf white pine, *Pinus strobus* 'Nana', ultimately reaches about 8 feet. *Picea pungens* 'Globosa', the popular bird's nest spruce (the growth actually resembles a bird's nest with a shallow depression in the center of the plant) will form a mound of gray-blue needles, some 18 inches high and 24 inches wide, after ten years of growth. Other conifers include *Juniperus horizontalis* 'Blue Tip', and the yellow-tipped form of the Swiss mountain pine, *Pinus mugo* 'Aurea'. Conifers are beautiful in the winter as they are the rest of the year. And they are adaptable too. This particular collection came to the Garden from Oregon, and some are over thirty years old. Surprisingly, most have adapted to the harsh St. Louis climate.

THE DESERT HOUSE

First built in 1914, the Desert House flanks the Climatron on the south, displaying plants from the arid regions of the world. Most of the cacti and succulents in the collection survive in desert situations where rainfall is less than ten inches a year (comparatively New York City has forty-three inches and London, England, has twenty-three inches).

Opposite, above:
The common jade plant, Crassula argentea, *popular for decades as a tough houseplant, blooms next to an organ-pipe cactus,* Lamaireocereus marginatus, *in the Desert House.*

Opposite, below:
Looking across the bizarre landscape of the Desert House. Many of these plants have been in the Garden's collection since before the First World War.

Pages 92–93:
An unknown hybrid of sea urchin cacti, Echinopsis *spp., carpets the sandy floor of the Desert House with their pleasant abstract design.*

Pages 94–95:
Like fingers in a many-fingered glove, a number of lace cacti, Mammillaria elongata, *surround a small clump of powder puff cactus,* M. bocasana. *Both plants originally came from Mexico.*

Walking down a sandy path with the twisted but stiff branches and leaves of desert plants on either side is indeed like visiting another world: round and squat barrel cactus resemble bowling balls covered with needles and are dwarfed by tall agaves with tentacle-like arms. The desert has a penchant for producing the very odd. The telegraph or boojum tree, *Idria columnaris,* is a bizarre species native to Baja, California, that resembles an upside-down carrot. The Garden's specimen is the largest one of its kind inside a conservatory. And there are a number of elephant-foot trees, *Beaucarnea recurvata,* which do indeed resemble the pachydermic umbrella stands that have been filled with fountains of curving 6-foot leaves. Although some of the plants like the American cacti and the African spurges belong to entirely different plant families, the similarity of their environments has led to their being strikingly familiar in appearance.

The Desert House will be renovated and rebuilt over the next few years. The new exterior design will match that of the Temperate House. Many of the plants will remain in situ and building will proceed around them. Some of the aloes and the agaves have been in the Garden's collection since before the 1904 Fair, and would perish if disturbed.

THE HARDY SUCCULENT GARDEN

Just east of the entrance to the Desert House is a garden of hardy succulents. These plants withstand the rigors of a Missouri winter with minimal protection, and if snow comes early and stays late, they do beautifully. Hardy cacti that are native to Missouri include *Opuntia macrorhiza,* a plant with innocent-looking soft spines that can easily pierce the human skin yet bloom in spring with enchanting yellow flowers. Other hardy species include Adam's needle, *Yucca filamentosa,* with its white flowers that are fragrant in the evenings, and the familiar low-growing succulents, the cobweb live-forevers, *Sempervivum arachnoideum* and *Sedum spurium.*

A number of desert annuals, plants that go through their entire life cycle in as short a time as three months, brighten the garden. Two native American plants include the lovely yellow desert marigold, *Baileya multiradiata,* and *Aster bigelovii* (*Machaeranthera bigelovii*), a fall-blooming perennial that bears purple or violet flower heads up to 2½ inches across with yellow centers, crowded with blossoms until well after the first frosts of autumn.

THE LEHMANN ROSE GARDEN

When leaving the Climatron complex and heading in the direction of the Japanese Garden is the sumptuously displayed Anne L. Lehmann Rose Garden.

Pages 96-97:
Outside the Desert House is the Hardy Succulent Garden, where plants with thick and fleshy leaves resist the perils of a drought and also survive the cold Missouri winters.

Opposite, above:
A view of the perfectly kept flower beds found in the Lehmann Rose Garden. Garden volunteers continually remove the spent blossoms from early spring into fall.

Opposite, below:
The Lehmann Rose Garden contains a large collection of magnificent roses, featuring many modern rose hybrids and old cultivars that have been popular over the years. Cultivars are plants cultivated by people, not by nature.

Above:
The spectacular beauty of the rose 'Pristine'.

Opposite:
A test rose for the All-America Rose Selections.

Overleaf:
North of the Lehmann Rose Garden is the Shapleigh Fountain, installed in 1974. Openings between three curved curtains of water invite the visitor to enter an outdoor room formed with the tall sprays.

Created in 1974, this rose garden covers about 1½ acres in size and is planned in curved beds that lead toward a gazebo situated at the head of the garden. In combination with the Gladney Rose Garden, there are slightly over two hundred rose cultivars in active growth—the Hybrid Teas, Floribundas, Grandifloras, Shrub Roses, and a few species.

In order to maintain nearly perfect rose gardens, the plants are deadheaded continuously right into fall. All the roses are fertilized in the spring when new growth begins and then every four to six weeks all summer long. In the late fall, the roses are mounded with six to eight inches of manure for winter protection, although they still suffer a ten percent loss.

It is here that horticulturists also test the newest rose cultivars in cooperation with the All-America Rose Selection and their yearly competition for the best new rose. Three of the newest Hybrid Tea selections on display are 'Tiffany', a bright, clear pink; 'Pristine', a soft white with a pink center, and 'Broadway', a bicolor that is mostly orange and red with a yellow center.

THE DEMONSTRATION VEGETABLE GARDEN

The Garden has always had a history of offering information to interested home gardeners. The Economic Garden with its model backyard vegetable and flower beds was popular for years, and endless pamphlets have been published with titles such as "Care of Home Plants" and "Growing Orchids from Seeds."

The Demonstration Vegetable Garden was started in 1982 and lies on the pathway between the Lehmann Rose Garden and the Japanese Garden. Beginning in early spring, the Garden's horticulturists plant sequential displays of the best vegetable varieties for the St. Louis region and many new cultivars just released to the market by commercial growers.

Further along the path is the site of the new Center for Home Gardening, an 8½-acre plot that will include twenty-three distinct gardens and plant displays, all designed to fit the needs of homeowners and amateur gardeners from around the area. A ten thousand-square-foot building will provide space for classes, instructional displays, and a library.

THE JAPANESE GARDEN

A Japanese garden should not be judged by Western rules. A Western garden is meant as a setting for plants and blossoms, a place where flowers rule and brilliant color is usually king. Japanese gardens excel in the wonders of natural beauty: a branch twisted by the wind, grasses bending in the snow, or a mighty rock etched by the hand of time. A

Opposite, above and below:
Raised beds in the Demonstration Vegetable Garden contain sequential displays of vegetables, including new varieties introduced by commercial growers.

tree that is beautiful when in flower is perceived as equally beautiful in the fall when its leaves turn to flame or in the winter when its bare branches are silhouetted against a muted sky.

In the world of the Japanese garden, more is not necessarily better. A plant with one bloom is not judged inferior to a plant with several blooms. The object of a Japanese garden is to refresh the mind. It is a garden to put the mind at peace through the quiet contemplation of natural beauty.

At the Missouri Botanical Garden the Japanese Garden is called *Seiwa-En,* a "garden of pure, clear harmony and peace." Dedicated in 1977, it is the largest such garden on the continent of North America, having fourteen acres of land that includes a 4½-acre lake, itself dotted with four islands. Professor Koichi Kawana, a native of Japan and lecturer in Japanese architecture, art, and landscape design at the University of California, Los Angeles, designed *Seiwa-En,* taking great care that the garden was truly authentic. Dr. Kawana visits the garden twice a year to observe plant growth, supervise tree pruning, and observe the progress of the garden.

Water in the Garden

In most Japanese gardens water is considered to be a primary element. For *Seiwa-En* every conceivable form of moving water found in Nature was reproduced and then given its own name. Where running water was not possible, ponds, stone basins, and wells were added. There are, for example, nineteen different kinds of water basins, each having a name derived from its shape.

Garden design in Japan was originally influenced by the Buddhist tradition from Korea and China—thus the lake at *Seiwa-En* is of an irregular shape, in keeping with the Oriental practice of designing the outline to represent the Chinese character for "mind." By walking along such an irregular course, the garden visitors experience constantly changing views, never really seeing quite the same thing twice.

Seiwa-En has a series of small waterfalls that symbolize mountain cascades. One, the blue boulder cascade, *Seigan-no-Taki,* is constructed with three great steps that symbolize heaven, earth, and man. *Cho-on-Baku* is the name for another waterfall that mimics the sound of the ocean's tides.

The Four Islands

Four islands rise from the lake. Tortoise Island and Crane Island are both named after Japanese symbols for longevity. These islands are deliberately isolated from the shore. A stone at one end of Tortoise Island represents the creature's head. The tail is represented by a rock at the other end, while additional stones represent the legs. On Crane Island, trees are planted diagonally to represent the wings of cranes in flight.

Opposite, above:
Here in the blue boulder cascade, Seigan-no-Taki, *water symbolizes heaven, man, and earth.*

Opposite, below:
Paradise Island, called Horaijimi *in Japanese mythology, is the symbolic center of* Seiwa-En. *Its three rocks represent everlasting happiness and immortality.*

Overleaf:
Looking toward Crane Island, one of four such islands rising from the four and one-half acre lake at Seiwa-En. *Trees are planted diagonally to represent the wings of cranes in flight.*

Formed by three large stones, Paradise Island is the symbolic center of the garden, representing everlasting happiness and immortality. Such stones are important features of Japanese gardens and are usually presented in combinations of three or five. In this environment, they are never viewed as just rocks, but are placed so as to interact and attract each other so perfectly that if one is moved, the entire composition is out of balance.

The innermost island is called Teahouse Island or Middle Island, and is connected to the mainland by two footbridges. There a traditional teahouse is surrounded by the tea garden or *Cha-niwa,* used by the *Cha-no-yu,* the ceremonial tea master. The garden includes a *Tsukubai* (water basin) to provide water for cleansing the hands. Like all the other features of the Japanese Garden, the teahouse has been authentically created. It was constructed in Nagano, Japan, according to ancient methods, then dismantled for shipment to St. Louis. A team of Japanese craftsmen from Matsumoto City traveled to the Garden in order to reconstruct the teahouse, and a Shinto priest performed the elaborate ceremony needed to signify the completion of the structure. The Island and its teahouse are meant to be viewed from a distance and neither is open to the public except for ceremonial occasions.

The Bridges of Seiwa-En

The bridges of *Seiwa-En* are not only intended to provide access from the shore to the islands, but to give visitors a place to view the changing reflections of the garden and the fish swimming in the lake.

Drum Bridge, *Taikobashi,* connects Teahouse Island with the northwest shore. The name refers to the circle completed by the bridge and its reflection in the water below. Made of redwood, it has been left unpainted to preserve its natural appearance. The bronze caps that top the four posts at each end of the bridge are called *giboshi.*

Crossing to the south shore from Teahouse Island is the Earthen Bridge, *Dobashi,* its base constructed from a row of logs on a timber framework and covered with a layer of earth. The outer edges of the bridge are planted with mondo grass, *Ophiopogon japonicus,* to control erosion.

Finally, along the southwest shore runs a unique zigzag bridge called *Yatsuhashi,* meaning eight bridges, and referring to a place in Mikawa Province, Japan, where a river branched out into eight channels in the tenth century. Each channel was crossed by a bridge surrounded by irises. At *Seiwa-En* soil-filled bins are filled with irises and placed on both sides of the bridge. In Japan, it was believed that evil spirits could only move in straight lines and not change course. If a person walked along a zig-zag bridge, a devil would be forced to pass and fall directly into the water.

Opposite:
The Drum Bridge connects Teahouse Island with the northwest mainland. Constructed of redwood, the bridge is left unpainted to preserve its natural appearance.

Overleaf:
Yatsuhashi, *the Zig-zag Bridge. Soil-filled bins of iris are placed on both sides of the bridge; even after the blooms are spent, the leaves make linear reflections in the water.*

In the Dry Garden, waving lines symbolize the ripple and wave patterns of water. The gardeners use special rakes with wide triangular tines to create the graceful designs.

The Dry Garden

If water is not present in part of a Japanese garden, it can be represented there symbolically. Using rocks, pebbles, gravel, and sand, a dried stream bed can be created. Here, along the southeastern shore of the lake, a dry garden of swirling patterns represents waves upon the ocean; the islands are mounds of trimmed evergreens.

The Lanterns

Throughout *Seiwa-En* there are fifteen traditional *Yukimi-gata* or Japanese stone lanterns, formerly used to guide the priests to the teahouse. Now they are usually called snow-viewing lanterns, so named because of the delicate look to the snow as it fell when seen by the lantern light. The lantern that stands at the entrance to the garden was originally from the 1904 World's Fair. Another lantern known as the *Rankei-gata* is placed so that its image is reflected in the water of the lake. And a third lantern stands on Teahouse Island, and is a gift from the city of Suwa in Japan.

The Four Seasons

Cherry trees, flowering crabapples, and azaleas are at their glory in late April and early May. Then peonies and iris bloom. During the summer, the lotus blossoms are reflected in the lake and then by September, waving plumes of the ornamental grasses, *Miscanthus sinensis* cultivars and *Pennisetum alopecuroides,* appear, followed by cascades of chrysanthemums that bloom well into November. Throughout the year pygmy bamboos, *Arundinaria pygmaea,* that only reach a height of ten inches, and a taller species, yellow-grove bamboo, *Phyllostachys aureosulcata,* add more of a feeling of the Orient to the garden.

Finally, in a Japanese garden, snow is considered as a flower and is called *sekka* or *toka.* In winter, the *Momiji,* or maple trees, now shorn of leaves, become stark sculptures to be viewed against the mantle of white, the mounds of dimpled snow outlined against the needles of the evergreens. And often, lights in the *Yukimi-gata* would appear to guide the *Cha-no-yu* teachers across the garden to the teahouse in the evening.

There are fifteen stone lanterns scattered about Seiwa-En. *Originally introduced into Japanese gardens to help guests find their way along the garden path in the evening, this particular lantern is a* Kasuga, *named for a Shinto deity.*

THE COHEN AMPHITHEATER

To the north of the Japanese Garden an outdoor amphitheater completed in 1983 assumes the shape of a bowl where visitors can sit on the lawn to enjoy music or theater under the sun or stars. Every year this amphitheater becomes a focal point for the Japanese Festival, an event that started in 1976 as a one-day celebration in June, coordinated

Japanese Koi fish usually swim lazily under the summer sun, but at feeding time their orange and reddish scales shimmer like beaten gold.

Golden apples on a Japanese flowering quince, Chaenomeles speciosa.

Overleaf:
Two Japanese flowering crab apples, Malus *'Brandywine', blossom with deep rose-pink double flowers in spring. They are surrounded by a sea of liriope. Rhododendron hybrids and yews,* Taxus × media, *are at the left.*

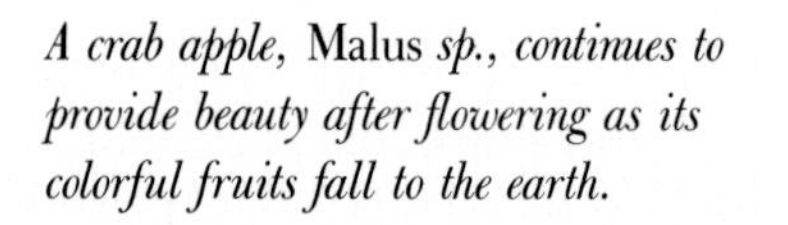

A crab apple, Malus *sp., continues to provide beauty after flowering as its colorful fruits fall to the earth.*

A bamboo hedge near the Cleveland Gate House. Known as the yellow-grove bamboo, Phyllostachys aureosulcata, *it is hardy to at least minus 10°F.*

Opposite, above:
Throughout the Japanese garden, liriope, Liriope muscari, *is used as an effective, low-maintenance ground cover. A member of the lily family, it blooms with small dark blue flowers in late summer.*

Opposite, below:
A traditional Japanese bamboo fence known as a Yotsume-gaki *serves as a background for lovely fall-flowering chrysanthemum cultivars from Japan.*

by the Japanese American Citizens League, and which now occurs ten days prior to and including Labor Day. Tens of thousands of people come to enjoy the art, food, culture, and music of Japan, and the amphitheater rings with the sound of the *taiko,* the traditional drums of Japan.

In the shade of the Woodland Garden, Hosta longipes *is used as a splendid groundcover. Many bell-shaped blossoms appear on 12-inch scapes or flowering stalks.*

ENGLISH WOODLAND GARDEN

East of the Japanese Garden, there is a teak bench on one of the side trails deep within the English Woodland Garden where one can sit and listen to the far-off voices of visitors in company with the scratching sound of squirrels spiraling around tree trunks and punctuated by the sharp sound of hickory and walnuts plummeting to the ground below. Under the canopy of trees, the air is cool even on a warm day in late summer. Fall-blooming hostas dot the shade with their white flowers; hardy begonias, *Begonia grandis,* bear fragrant flesh-colored blossoms; and off to one side, an unusual American wild flower, the zigzag spiderwort, *Tradescantia subaspera,* holds deep blue flowers on stems that look like bolts of lightning drawn by a child. House sparrows rustle through the leaves in search of small insects and seeds.

This garden was designed by John Elsley, a horticulturist trained at Kew Gardens, England, and developed as the Garden's project for the nation's Bicentennial in 1976. About 3 acres in size, it consists of three vegetative layers: an upper tree zone or canopy, a middle layer devoted to shrubs, and the lower layer of herbaceous plants. Here and there sun pierces to the forest floor in glades, dotting the woodland floor with pools of light. There are more than six hundred feet of wood-chip pathways scattered about the forest.

In the early spring the snowflakes, *Leucojum vernum,* and the winter aconites, *Eranthus hyemalis,* bloom, followed by a host of dogwoods, *Cornus florida.* The collection of wild flowers and perennials could be termed a floral potpourri since it mixes *Bergenia ciliata; Acanthus mollis;* witch hazel, *Hamamelis* spp.; Solomon's seal, both *Polygonatum commutatum* and other species; hostas; bugbanes, species of *Cimicifuga;* Peegee hydrangeas, *Hydrangea paniculata;* a number of rhododendron hybrids; and autumn crocus, *Colchicum speciosum,* that bloom in mid-September. There is a magnificent bed of variegated Solomon seal and blackberry lily, *Belamcanda chinensis.* A large patch of horsetail, *Equisetum hyemale*—an interesting living fossil that is usually not found in most gardens—fills a spot in deep shade and the lack of light seems to keep it in check.

Presently terrestrial orchids from a collection in Iowa are joining the other wild flowers in the English Woodland Garden and will include the large yellow lady's-slipper, *Cypripedium calceolus* var. *parviflorum;* the small white lady's-slipper, *C. candidum;* and the showy lady's-slipper, *C. reginae.*

Opposite:
A bed of coneflowers, Rudbeckia fulgida *var.* sullivantii, *borders one of the many paths in the English Woodland Garden. Trumpet creeper vines,* Campsis × tagliabuana *'Madame Galen', clamber over a dead tree in the background.*

A rare and endangered coneflower, Rudbeckia tennesseensis, *is grown as part of the Center for Plant Conservation project.*

The Endangered Species

Throughout the Garden endangered plant species are marked with the symbol of a panda. Their presence is owed to the Center for Plant Conservation Project, a national effort dedicated to protecting rare species through cultivation at botanical gardens throughout the country.

The English Woodland Garden includes pondberry *Lindera melissifolia,* one of the rarest species in the country. Today pondberry is only found in about eight places, one of them in southern Missouri, the Sand Pond Natural Area. The scarcity of the plant, combined with threats to the few populations left and poor reproduction in the wild, resulted in its becoming listed as endangered in 1986 under the Endangered Species Act. The pondberry is related to the much more common spicebush, *L. benzoin,* although it has a different odor, droopier leaves, and bigger fruits on stouter and longer stalks.

The Missouri population was first discovered by Julian Steyermark, a Garden botanist, in 1948 but then lost until 1979. That year another botanist was attempting to relocate the lost colony when his car broke down in a heavy rain storm. He went to a nearby farm house to ask for help and in conversation with the farmer discovered that the pondberry patch was on his farm.

Members of the Garden staff report to the Center for Plant Conservation on endangered species. Based on their findings, other examples at the Garden include Lucy Braun's eupatorium, *Eupatorium luciae-brauniae.* It is closely related to several familiar wild flowers: Joe-Pye weed, *E. purpureum,* and white snakeroot, *E. rugosum.* The latter, when eaten by cows, was the midwestern source of the deadly milk sickness that plagued early settlers and is reputed to have killed Abraham Lincoln's mother.

Lucy Braun's eupatorium is a shy member of the genus, living under rock overhangs where its greatest danger comes from trampling by rock climbers and from covering by silt resulting from nearby lumbering.

There is also a small patch of running buffalo clover, *Trifolium stoloniferum.* First collected near St. Louis, this clover was once widespread and fairly common in the Midwest. A survey of writings left by early settlers in Kentucky turned up repeated references to it. Some botanists believe that this plant was so closely associated with buffaloes that as these animals perished so did the plant. Presently only four populations are known, found in Kentucky, Indiana, and West Virginia. A fifth patch, also in West Virginia, was destroyed in 1986.

Birds in the Garden

"I knew the stars, the flowers, and the birds," wrote John Millington Synge, for like most gardeners he recognized that a garden without birds would be a decidedly strange place to visit.

Every year over one hundred species of birds can be seen at the Garden. Surrounded by the city and all its noise, speed, and cars, the

Garden becomes a green oasis, not only to the birds that nest in the area but to those that pause while in flight to other climes. Thirty-one species of warblers, six species of vireos, six of the thrush clan, six different woodpeckers, plus various orioles, tanagers, cuckoos, swallows, flycatchers, and finches visit every year.

The Japanese Garden provides both insects and water for swallows and green herons live in the clumps of lotus in the lagoon. Redwing-blackbirds nest along the shore and warbling vireos can be heard singing high in the cypress trees, while in the early evening the nighthawks come out to feed.

The English Woodland Garden is home to warblers, vireos, and wood thrushes. Sparrows rustle in the leaves and nesting robins, cardinals, blue jays, and blackbirds fly about in the canopy of trees.

The Mausoleum Garden is an especially fine spot to sight birds as the tall oaks and yellow buckeye offer a safe spot to nest and a fine lookout. Migrating warblers, vireos, cuckoos, tanagers, orioles, and thrushes are there in the spring and the fall while robins, cardinals, blue jays, and a number of sparrow species nest in the holly trees and the low bushes.

Spring and fall are major times for bird migrations and Garden visitors might hear the sharp cry of a hawk as it wheels through the sky. Sharp-shinned hawks, Cooper's hawks, and red-tailed hawks are common. Even a peregrine falcon was sighted doing a diving and soaring display over the steps at Ridgway Center.

The Garden sponsors birdwalks in the spring, when veteran birdwatchers lead visitors on tours and help with identification.

Sassafras, Sassafras albidum. *In* The Natural history of Carolina, Florida and the Bahama Islands, *vol. 1 by Mark Catesby (1754).*

THE HERB GARDEN

Directly behind Henry Shaw's Tower Grove House is the Herb Garden, an area of about two thousand square feet, enclosed by a painted black iron fence in the Victorian style.

When walking through the west gate of the garden one passes between two six-foot-high ornamental grasses, but these are not any of the common varieties but instead are Vetiver grass, *Vetiveria zizanioides*, or the tropical Khus-Khus, having stiff branches at the base, opening to graceful bends toward the top. Each year one plant is divided and kept in the greenhouse until the following summer, and the other has its roots dried and kept for the fragrant oils they produce.

Besides most of the typical culinary and medicinal herbs, the Herb Garden contains more unusual plants such as a patch of the saffron crocus, *Crocus sativus*, grown commercially to color and flavor foods, and *Carthamus tinctorius*, the annual used to produce safflower oil.

Within the garden is the *Child Sundial*, a lead statue of a child daydreaming with a sundial at ground level and mounted in a bed of creeping thyme, *Thymus vulgaris*.

Overleaf:
A view of the Herb Garden, located behind the Tower Grove House. The display consists of small individual beds of both culinary and medicinal herbs.

Tower Grove House was the country home of Henry Shaw and took its name from the grove of sassafras trees now in the Mausoleum Garden. The interior is open to the public and is furnished in the style of its day.

Henry Moore said of his bronze Two Piece Reclining Figure II, *made in 1960, "You don't expect it to be a naturalistic figure; therefore, you can more justifiably make it like a landscape or a rock." It is situated north of the Lehmann Building.*

Juno *looks toward the Museum Building. The figure is a copy of the* Farnese Juno, *fifth century B.C., executed in white marble by Carlo Nicoli in 1885 and brought to the Garden in 1887.*

In the Herb Garden, a patch of ornamental peppers, Capsicum annuum *'Thai Hot', are not grown for the flowers but for the pretty, colorful fruits and attractive foliage.*

Above, left:
Growing luxuriously in the summer heat, bunches of lavender cotton, Santolina chamaecyparissus, *surround a cultivar of the common garden sage,* Salvia officinalis *'Tricolor'.*

On the west side of the house is the bronze *Fountain Angel* by Raffaello Romanelli (1856–1928), standing eighty-one inches high from wing tip to base. It originally stood in front of a marble column at the 1904 World's Fair and was restored and installed at Tower Grove House in 1975.

THE KAESER MEMORIAL MAZE

Every botanical garden should have a living maze. In 1986, the Kaeser Memorial Maze was begun on the site of an old greenhouse southeast of Tower Grove House. Visitors wander through a labyrinth of yew hedges, *Taxus* 'Hicksi', in and about a Clematis-clad pergola at the center of the maze. Because the maze is still young and the bed depressed two and one-half feet, outsiders can at present watch the progress of stalwart explorers (usually that is, children) as they wander the graveled pathways.

MAUSOLEUM GROUNDS

In 1882, Henry Shaw was eighty-two years old. He asked his architect and friend, George Barnett, to design an octagonal mausoleum of white Missouri limestone to be placed in a grove of sassafras trees, *Sassafras albidum,* directly across from Tower Grove House and to serve as Shaw's final resting place.

Overleaf:
The Kaeser Memorial Maze sits within the triangle formed by Tower Grove House, the Museum Building, and Henry Shaw's townhouse, now the Administration Building.

Sassafras near Shaw's Tower Grove House.

Opposite:
Henry Shaw's Mausoleum stands at the rear of the Mausoleum Grounds. Circular beds of fancy-leaved caladiums, Caladium *'White Wings', separate, at right, a planting of liriope and at left a species of euonymus. The trees are all native oaks.*

In 1885, Shaw changed his mind, deciding that a granite tomb be placed in the sassafras grove. The limestone building was moved further north and now stands on the walkway that passes the north gate to the Mausoleum. Shaw ordered a white marble statue of a partially draped seated woman writing on a shield, called *The Victory of Science over Ignorance* by Carlo Nicoli of Italy, to be placed in the limestone mausoleum. The statue looks out on the Knolls, and the unused tomb is set off by marvelous clumps of ornamental grass, *Miscanthus sinensis* 'Variegatus'.

The granite mausoleum, also designed by Barnett, is an octagon of pink granite with a domed roof of patinaed copper topped by a cross. Shaw's marble effigy—a reclining figure carved by Baron Ferdinand von Miller of Germany—lies within, surrounded by tall windows that are edged with colored shards of Victorian leaded-glass.

The Mausoleum grounds are an oval of ground just under an acre in size; a black iron fence with gates at either end encloses the long diameter. During the summer, the four black iron Grecian urns atop the gateposts are filled with cascading spider plants, *Chlorophytum comosum.* It is a pleasant garden in which to sit, especially on a hot summer's day, under the shade of five species of oak, *Quercus* spp.; yellow buckeyes, *Aesculus octandra,* that eventually tower to ninety feet; and the sassafras, direct descendents of the originals present at Shaw's death, in 1889.

In the spring the mausoleum is surrounded by a vast field of Spanish bluebells, *Endymion hispanicus.* During the summer, the grounds are carpeted with lilyturf, *Liriope muscari.* Many of the walkways are edged with white caladiums, *Caladium* 'White Wings', and are surrounded with beds of Baltic ivy, *Hedera* 'Baltica', a cultivar introduced to this country by Dr. Edgar Anderson, the Garden's director in the mid-1950s. Additional beds of bracken fern, *Pteridium aquilinum,* and a number of hostas complete the groundcovers.

The land that Henry Shaw developed as the Garden was not the forest primeval but rather grassland and meadow. Early records on the Garden's development are not complete, but probably the only trees in existence today that go back to the birth of the city are the roots of the grove of sassafras trees in the Mausoleum.

There is a gigantic yellowwood tree, *Cladrastis kentuckea,* on the east side of the Garden near Flora Gate. Obviously of great age, the tree resembles a medieval beggar in a Rembrandt drawing with its heavy gnarled limbs projecting on all sides of the massive trunk, held up by a network of large wooden crutches.

To the east of the Cohen Amphitheater, on the west side of the Garden, are three very old bald cypresses, *Taxodium distichum,* that go back to the Garden's beginnings. And to the west of Tower Grove House is a magnificent Wych or Scotch elm, *Ulmus glabra.* Finally, between the Mausoleum grounds and the Lehmann Rose Garden is one old Ginkgo tree, *Ginkgo biloba,* with another in front of the Temperate House.

The statue of Victory *stands within Shaw's first mausoleum at the edge of the Knolls and is a copy of a piece in the Pitti Palace, Florence. Nicoli made this in 1885, the same year he sculpted* Juno.

Overleaf, left:
At the edge of the Knolls, clumps of the shrubby bottlebrush buckeye or dwarf horse chestnut, Aesculus parviflora, *are shaded by several species of oak.*

Overleaf, right:
Osage oranges, Maclura pomifera, *west of the Climatron. These trees are remnants of a row that lined the private drive to Henry Shaw's Tower Grove House. In spite of their common name, Osage oranges are related to mulberries. They are not native to Missouri, but were brought from the west by French colonials. Lewis and Clark collected cuttings in a St. Louis garden in 1804 and sent them to Thomas Jefferson, who is responsible for having them introduced into horticulture and for having them described to science.*

A stabile by Alexander Calder, entitled The Tree, *stands within a grove of black walnut trees,* Juglans nigra. *A lone bitternut hickory,* Carya cordiformis, *grows just behind the sculpture.*

Prairie wild flowers blooming near the Knolls.

The gnarled-looking spindle tree, Euonymus bungeana, *is a shrub or tree from Japan, and is surprisingly hardy in the St. Louis climate.*

The oak-leaf hydrangea, Hydrangea quercifolia, *is used as a background shrub in the Jenkins Daylily Garden. A native American plant, it was discovered in 1791 and thrives in deep shade.*

THE JENKINS DAYLILY GARDEN

Next to the rose and the hosta, the daylily, species and cultivars of *Hemerocallis,* is probably the most popular perennial in America today. Almost impervious to pests, very resistant to drought, and tolerant of most garden sites, daylilies flower in all colors except white and blue, from the beginning of June on into October. Each blossom lasts just one day, but new buds are produced over a long period, guaranteeing weeks of bloom from each plant.

Although only dedicated formally in 1988, the Jenkins Daylily Garden contains 600 varieties of daylilies. With beds on both sides of the walkway that leads from Tower Grove House to Flora Gate, the daylilies cover nearly an acre. New plants continue to come into the garden from regional breeders located all over, from Pennsylvania to Florida and California. There will be about 900 cultivars in the collection by the time the beds are established and finished.

Most daylilies in the garden are usually between two and three feet in height when in flower, but *Hemerocallis altissima,* a species from China that blossoms in late summer and autumn, has buds and flowers on stems, or scapes, often six feet tall. Yet *H. minor,* the dwarf yellow daylily originally from Eastern Siberia and Japan, is under twenty inches. Cultivars such as the new 'Stella d'Oro' or 'Everblooming Doll' will produce flowers over most of the growing season.

The garden's daylilies are mulched in the fall and given a bit of superphosphate. All the spent flowers and the scapes are removed after bloom. By following this regimen, the leaves stay greener, thus turning brown less quickly, and the energy that would ordinarily go into producing unwanted seed leads instead to bigger and better flowers the following year.

The beds are heightened with hydrangeas, including the autumn-blooming *H. paniculata,* in company with clumps of ornamental grasses.

The Prairie Garden consists of native American wild flowers. Here tall blazing stars, Liatris pycnostachya, *hover above a lone clump of black-eyed Susans,* Rudbeckia hirta, *which bloom in late August and September.*

GOODMAN IRIS GARDEN

Iris gardens burst open in mid-spring and then continue blossoming with such abandon that they will steal the show from all other flowers in the neighborhood, and the Goodman Iris Garden is no exception. When the iris beds are at their peak on or about May 15, and twelve hundred blossoms in all the colors of the rainbow open to a late spring sky, they are brilliantly beautiful. The colors are truly spectacular and so popular that between twelve and fifteen thousand people will come to this garden for the display.

Although there have always been iris beds at the Garden, *Iris* species and cultivars, it was in 1984 that the spot near the Milles Sculpture Garden was designated for the Goodman Iris Garden. At that time beds in the shape of a half-moon were designed—the fan-like foliage

A cultivated or bearded iris, Iris × germanica, *is named for the "beard" or pattern of hairs found at the top of the fall, the term for the drooping petals of the flower.*

Two modern daylily cultivars, from the Jenkins Daylily Garden: 'My Funny Valentine' and 'Mary Todd'.

The semi-circular beds of the Goodman Iris garden are brilliant in May, when this outstanding collection of both species and cultivars is in full bloom.

An unknown cultivar of Siberian iris, Iris siberica, *not only is attractive in bloom but the leaves make a fine garden accent.*

and frothy petals of iris are best displayed in serpentine beds—and the transplanting began. Individual beds are five feet wide with four-foot aisles in between.

This particular garden is even beautiful out of season. It is so well groomed that almost every burned tip on every leaf has been carefully cut off so that stiff fans with blunted leaves become a stunning layered pattern of bluish-gray.

Back in 1981 the American Iris Society planned to hold their big convention in St. Louis. To prepare for this, iris hybridizers sent plants from all over the world two years ahead of time. These iris were then placed in the gardens of various members throughout the area so that when the big day arrived, everything would be acclimatized and in perfect shape. One plant of every cultivar slated for the convention went to the Garden. At this time the Iris Garden contains over fifteen hundred iris cultivars plus about twenty species.

According to a plaque in the garden, the French heraldic symbol of the fleur-de-lis is thought to have started with the hybrid German iris, *Iris × germanica.* The original flower inspiration is believed to be the European and North African yellow waterflag, *I. pseudacorus.* According to legend, this iris helped save a French army in the sixth century from its enemy that it had cornered in a bend in the Rhine River. The French king recognized a potential ford in the river by spotting the iris in the water and directed his army across the shallow spot to safety. The grateful king chose the iris as the symbol of the Royal Family. Later King Louis VII maintained the iris as the symbol of the Crusades. At this time the flower was known as the fleur-de-Lis which became the present name over the years.

The Unicorn Spirit

Across from the iris beds is a bronze sculpture called *Unicorn Spirit,* by John Goodman, son of the iris garden's namesake. The helical form is 65 by 15 inches and is based on the double helix of the DNA molecule, which contains the genetic code for all living things and the spiraling of a unicorn's horn. The inside surfaces are inscribed with the forms of the fleur-de-lis.

Directly across from the Iris Garden is a bronze helix entitled, Unicorn spirit, *by John Goodman.*

THE KNOLLS

The rolling greensward called the Knolls lies between the Milles Sculpture Garden and the Mausoleum. It was designed by John Noyes, the Garden's resident landscape architect in 1914, to relieve the otherwise flat expanse of the Garden. This "return to nature" was also a turn-of-the-century rebellion against the artificial contrivances and strict formalities of Victorian landscaping, a trend that was just as popular in America as it was in England.

The velvety lawn is planted with a variety of stately trees. Just off-center is a large linden tree, *Tilia* spp., and when standing at the water lily pool and looking south toward the Knolls on the near left edge is a grove of Chinese Empress trees, *Paulownia tomentosa,* which bloom in spring with exotic and fragrant pale violet flowers. In the fall the tropical-like leaves of cutleaf staghorn sumac, *Rhus typhina* var. *laciniata,* turn to brilliant orange, a color that literally pops against the bright green grass.

The magnificent lawn known as the Knolls. From left are a winged eunonymus, Euonymus alata; *a young* Eucommia ulmoides; *and just behind that tree, a lace-leaved sumac,* Rhus typhina *'Laciniata'. The Museum building can be seen at far right.*

The Dry Stream Garden.

The Dry Stream Garden contains many vigorous perennials that prefer a place with full sun exposure. From left, an ornamental grass, Miscanthus sinensis *'Variegatus', is in stark contrast to the daylilies,* Hemerocallis *'So Lovely' at right, and the purple loosestrife,* Lythrum salicaria.

THE DRY STREAM GARDEN

Started in 1982, this garden is part of the Knolls. There a stream bed is constructed of white pebbles somewhat in the manner of a Japanese garden but without a hint of formality. The meandering "stream" is edged here and there with very large limestone rocks. Vast beds of daylilies; *Iris pseudacorus; I. virginica;* fountain-like mounds of the pendulous wood sedge, *Carex pendula;* and large clumps of *Miscanthus sinensis* 'Variegatus' highlight the plantings. A small pond at one end is full of hardy water lilies and various other aquatic plants.

THE LINNEAN HOUSE

There is a marvelous photograph in the archives, c. 1882–85, of the Garden that pictures a young lady in front of the Linnean House playing a violin—with all the implied fervor of a diva at Carnegie Hall—only instead of standing at stage center, she is seen posing in the middle of a giant water lily leaf, the tropically romantic *Victoria cruziana* (a species that will take cooler water temperatures than *V. amazonica*). The photograph seems to sum up the Victorian fascination for the bizarre and unique.

The Linnean House opened in 1882 and is the oldest continually operating display greenhouse in the United States. It was built to house a collection of camellias and to store palm trees over the hard Missouri winters. Henry Shaw named the building in honor of Carl Linnaeus, the eighteenth-century Swedish botanist who devised the present method of naming plants and animals. Three marble busts by the American sculptor, Howard Kretschmar, adorn the building's pediment: Linnaeus; Thomas Nuttall, the most famous nineteenth-century botanical explorer of the American West; and Asa Gray, the leading American botanist of that time.

The camellias begin to bloom in early November with the flowers of *Camellia sasanqua,* followed by the main display consisting of *C. japonica,* and ending with late-flowering types in March. Depending on the weather—camellias do not tolerate heat well—the Garden can have flowers well into April. The greenhouse uses artificial heat but in sparing amounts since the plants can easily tolerate chilly temperatures.

The spring display of primroses, *Primula malacoides,* with their white and pink flowers in bloom at ground level on either side of the brick paths can warm a heart jaded by winter. As the weather turns warmer they too begin to fade, but by that time attention has turned to the out-of-doors. Then by midsummer, the house is again in bloom with Lily-of-the-Nile, *Agapanthus africanus;* spider lilies, *Lycorus radiata;* and mounds of impatiens backed by the glowing leaves of angel wings, *Caladium × hortulanum.*

The Linnean House reflected in a pool full of tropical water lilies. This is the only greenhouse in the Garden dating from Shaw's lifetime. The busts over the main entrance are, from left, Thomas Nuttall, Carl Linnaeus, and Asa Gray.

Interior of the Linnean House.

The Cohen Court

On the east end of the Linnean House is a small brick-paved courtyard, often called the Yellow Garden. Here in the summer the flowers consist of a grouping of white and lemon swirl lantana bushes, *Lantana camara,* backed by mounds of Chinese pennisetum, *Pennisetum alopecuroides,* in turn flanked by a clipped yew hedge. A teak garden bench (of naturally weathered wood) is placed for contemplation. Beds of variegated Saint Augustine grass, *Stenotaphrum secundatum* 'Variegatum', are topped with gray-leaved euryops, *Euryops pectinatus,* and stand next to huge banks of *Melampodium tadulosum* 'Medalion' and beds of *Hosta* 'Francis William'. Four paper-barked maples, *Acer griseum,* and more Chinese pennisetum complete the arrangement. On

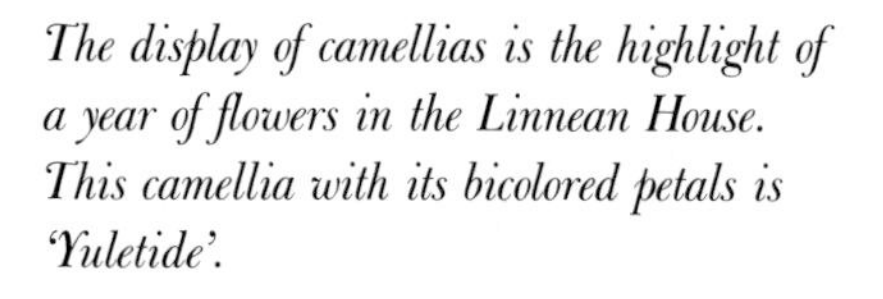

The display of camellias is the highlight of a year of flowers in the Linnean House. This camellia with its bicolored petals is 'Yuletide'.

A swallowtail butterfly alights on the blossoms of lantana.

Opposite, above:
Primroses and camellia blossoms.

Opposite, below:
The Three Graces *(1956) by the German sculptor, Gerhard Marcks, stand above the water of a reflecting pool near the Linnean House. Marcks, who was born in Berlin, was associated with the Bauhaus during the 1920s.*

The floating leaves of the Victoria water lily, Victoria amazonica, *are many feet in diameter. The flowers are creamy white to pink, depending on the time of day, and have an odor of pineapples.*

Many water lily cultivars in the Garden's collections bloom during the day and were introduced by one-time superintendent, George Pring. The flower is a hardy type, Nymphaea *'Pink Platter'.*

Opposite:
The Arbor between the Linnean House and the Scented Garden offers a peaceful place to escape the sun of a hot summer's day.

Looking across the well-tended beds of the Gladney Rose Garden. All the roses are carefully watered once or twice a week with overhead sprinklers fixed to poles that are set throughout the beds.

either side of the door to the greenhouse are Japanese anemones, *Anemone × hybrida,* of the palest shell pink.

The front door of the Linnean House looks out on pools full of water lilies and aquatic plants that front the flowering borders of the Swift Family Garden, where old-fashioned gas-lit lamp posts are festooned with giant balls of verbena that cast shadows on beds of maiden grass, *Miscanthus sinensis* 'Gracillimus'; rudbeckias; dusty miller, *Senecio cineraria* 'Snowstorm'; trailing lantana, *Lantana montevidensis;* and in the early fall, more huge clumps of Japanese anemones. Beyond one can see the Gladney Rose Garden and the Sculpture Garden.

GLADNEY ROSE GARDEN

The Gladney Rose Garden is everything that one wants of such a garden: a formal, old-fashioned white fence with moon-shaped arbors, covered with many varieties of climbing roses. It surrounds a quarter-acre garden. The Gladney was started in 1917, and circular beds of hybrid teas, floribundas, and grandiflora roses bloom all summer and on into fall, now reflected in a pool installed in the 1960s.

The broad pathways of the rose garden invite the visitor to enjoy a quiet stroll among the roses, or rest on a white garden bench.

'Pristine', a modern rose cultivar in the Gladney Garden.

Overleaf:
Designed in the shape of a giant wheel, the Gladney Rose Garden is surrounded by a white wooden fence, its arbors offering support for old-fashioned climbing roses.

The raised beds in the Scented Garden give visitors confined to wheelchairs an opportunity to easily view the plantings. The labels are all in braille, making this garden accessible to the blind.

Within the Scented Garden a number of bells crafted by Paolo Soleri in 1986 are suspended from a bronze tree-like structure over eight feet high, designed by William Severson and Vernon Gross. The slightest breeze will produce the bell tones.

THE SCENTED GARDEN

Following along one of the paths leaving the Linnean House, takes the visitor to the Scented Garden, opened in 1983. Designed to accommodate the blind, the labels are marked in braille, and the plants are in raised beds, accessible for wheelchairs. The species included are noted either for their interesting textures or delightful and tantalizing perfumes.

Bluebeard, *Caryopteris* × *clandonensis* 'Blue Mist', is not only charming to see with its airy froth of light blue flowers, but the leaves have a lovely smell of mint. Chinese chives, *Allium tuberosum,* stand straight as soldiers with bunches of white star-like blossoms on top of stiff green stems; they have a mild garlic flavor. Wall flowers, *Erysimum* spp., grow in the center of one bed, a rush of silvery leaves combined with a new ornamental basil, *Ocimum* 'Spicy Globe', that mounds beautifully and has a wonderful scent. *Achillea* 'Crimson Beauty'; French lavender, *Lavandula dentata;* a host of scented geraniums, *Pelargonium* spp., include lemon, apple, and nutmeg; and, of course, lamb's-ears, *Stachys byzantina,* a plant that looks exactly like its namesake, are all part of the Scented Garden.

With the slightest wind, *Bell Chimes,* a bronze sculpture that holds a host of Solari bells, vibrates with sound.

Sweet fern with the odor of sun-washed hills in summer, *Comptonia peregrina* var. *asplenifolia,* is next to the lace shrub from Japan, *Stephanandra incisa* 'Crispa'. Just at the far edge of the garden is a tree peony with the marvelous name of 'Furnace for Making Pills of Immortality'.

While only six years old, the Hosta Garden has the look and feel of a much older installation, because the stalwart plants thrive beneath the shade of the Garden's stately trees. Spring bulbs flower before the hostas begin their growth.

THE HOSTA GARDEN

Hostas, next to daylilies, are close to being a perfect perennial providing both lush foliage and lovely flowers, and they include a number of cultivars that will grow happily in shade.

The Hosta Garden is new, having opened in 1983, but already the plants have the look of a garden settled some years before. Just south of the Scented Garden, ten species that include over forty cultivars are carefully planted in carefully matched leaf patterns.

Hosta tardifolia, blooming in September with whitish-blue flowers, and *H. lancifolia,* with smaller dark green leaves, are lined up in front of a non-vining clematis, *Clematis recta* 'Recta Purpurea', in turn planted in front of a group of oak-leaf hydrangea, *Hydrangea quercifolia.* The late-blooming *Astilbe chinensis* and the new *Houttuynia cordata* 'Variegata' from the mountains of Java, Nepal, and Japan round out the plant combinations.

Overleaf, left:
The main axis between the Flora Gate and the Climatron consists of three reflecting pools that contain both water lilies and the Milles Sculpture Garden.

Overleaf, right:
Seven sculptures by Carl Milles are found within the reflecting pools. Above, three Angel Musicians *each stand on one foot atop sheer carnelian granite columns and cavort above the Garden's collection of tropical water lilies. They are 8 feet high, and were cast by the Swedish sculptor in 1950.*

FLORA GATE AND THE SCULPTURE GARDEN

A limestone wall along the eastern side of the Garden is bisected by Flora Gate. The gate and the wall were constructed in 1858 and served as the Garden's entrance until the opening of Ridgway Center in 1982. On the avenue that extends from the gate to the Climatron, three reflecting pools comprise the Milles Sculpture Garden. It is lined on either side by bald cypresses. Beds on either side of the water lily ponds

Two Girls Dancing *by Milles dates from 1917. The sculptor worked for a time in the studio of Auguste Rodin and critics have noted Rodin's influence.*

contain a new ornamental pepper, *Capsicum annuum* 'Dwarf Jigsaw'; *Pelargonium* 'Orbit Series Pink'; an ornamental basil, *Ocimum* 'Purple Ruffles'; bordered on either side by a variegated Jacob's-coat, *Acalypha wilkesiana* 'Variegata'. Standards of *Hibiscus rosa-sinensis* 'Cooperi' stand in pots.

Three pools were constructed in 1913. The central and circular pool is home to the Amazon water lilies, *Victoria cruziana* and *Victoria* 'Longwood Hybrid', which often produces leaves up to six feet wide. *Three Angel Musicians* by Carl Milles (1875–1955) were installed in 1988. The sculptures are sand-mold casts in bronze from the original models preserved at Millesgården, the sculptor's estate near Stockholm.

Two other rectangular pools hold the Garden's collection of tropical water lily hybrids, many developed by George Pring, a leading horticulturist and superintendent of the Garden with a career that spanned 1906 to 1963. Pring was responsible for introducing a day-blooming water lily and developing a number of currently available cultivars.

The pool near the Climatron holds *Two Girls Dancing*, 1914–17, the earliest Milles work in the Garden's collection. *Sun Glitter* and two *Orpheus Fountain Figures* rise above the water of the pool that fronts Flora Gate.

THE SAMUELS BULB GARDEN

To the north of the Sculpture Garden, the Samuels Bulb Garden is a spring tapestry of tulips, hyacinths, and daffodils, with the majority being tulips. There are also a number of species tulips including the water lily tulip, *Tulipa kaufmanniana*, with its pale yellow petals and a red-marked center of yellow, and *T. greigii*, that not only bears spectacular orange-red flowers, but has leaves that are distinctly mottled and striped with red and purple. Occasionally there are changes in the basic display because of unexpected color clashes, and hyacinths are often replaced because of the rigors of the winters.

Small low-growing shrubs planted throughout the garden give added interest when the bulbs are out of flower. The low-growing *Spiraea* 'Lime Mound' share space with various viburnums, *Viburnum* spp. and hybrids, which are consistently pruned so as not to overpower the bulbs.

Opposite:
Springtime in the Samuels Bulb Garden. Grape hyacinth, Muscari, *'Blue Spike', are in the foreground; behind them are Narcissus 'Duke of Windsor'.*

Overleaf:
The fiery hue of Tulipa fosteriana *'Juan' in the Bulb Garden.*

THE SHAW ARBORETUM

When Shaw began his garden it was located in the suburbs of St. Louis. By 1923, the city surrounded the Garden on all sides. Coal-burning industrial plants, railroad engines, and stoves in neighboring homes continually spewed forth thick clouds of black smoke that cut down on available sunlight and settled like a shroud over the Garden.

The willow pond at Shaw's Arboretum (1892), originally part of the Garden main grounds.

Whitetail deer at the Shaw Arboretum, Gray Summit, some forty miles southwest of St. Louis.

The woods in the spring at the Arboretum.

Growth was stunted and a number of the trees, shrubs, and herbaceous plants were killed. Particularly harmful was sulfuric acid, just one of the toxic elements found in the fumes.

In 1925, five abandoned farms totaling 1,650 acres were purchased near Gray Summit, situated on the Ozark Plateau, about forty miles southwest of St. Louis. Greenhouses were constructed, and the Garden's orchid collection was temporarily moved out to the country. A fifty-five acre Pinetum was developed with conifers brought from all parts of the world. In fact, there was even some talk about moving the Garden away from the scourge of industrial pollution. But by the mid-1930s, the city had taken steps to curb the smoke and the Garden gave up thoughts of leaving St. Louis. Today the Shaw Arboretum covers 2,400 acres of woods and prairies.

A 300-acre woodland wild flower habitat, with seven miles of trails, was started in the 1930s. Every spring, two hundred and fifty wild flowers bloom in succession beginning with bloodroot, *Sanguinaria canadensis,* in March, followed by a carpet of Virginia bluebells, *Mertensia virginica,* in April, to the rare, bell-flowered Fremont's leather flower, *Clematis fremontii,* in early May.

The Pinetum is not only known for the conifer collection, but it too, in spring, becomes a huge, glorious field of daffodils, carpeting the hills with gold.

More recent is the 74-acre Experimental Prairie, started in 1980 and now slowly maturing. Since the beginning, more than 38,000 greenhouse seedlings of 90 species of grasses and wild flowers have been added to the 113 species already found on the site for a total of 203 prairie plant species. Tall blazing stars, *Liatris aspera,* bloom with magenta-colored flower spikes in late July; many of the taller grasses reach a height of 10 feet by August, followed by drifts of pale purple coneflowers, *Echinacea purpurea.*

THE FUTURE OF THE GARDEN

Being in a garden or a park is as close as many people ever come to walking with and seeing something outside of their limited world that is full of life and beauty. A botanical garden becomes something very special in this sense, and when it instructs but never, never lectures, it means even more, becoming a place that leads the way to understanding by the example it sets. The Missouri Botanical Garden is such a place. Its first one hundred and thirty years have been grand. The next century, it is hoped, will be even grander. Goethe said, "To know of someone here and there whom we accord with, who is living on with us, even in silence—this makes our earthly ball a peopled garden." You can meet such people—and in such a place—in St. Louis.

RESEARCH: AT HOME AND IN THE FIELD

The Museum Building was the original home for Shaw's collections and designed by George I. Barnett, the architect of Tower Grove House.

The scientist values research by the size of its contribution to that huge, logically articulated structure of ideas which is already, though not half built, the most glorious accomplishment of mankind.

—Sir Peter Brian Medawar, *The Art of the Soluble* (1967)

Along the eastern edge of the Garden is the museum that Shaw constructed to house his herbarium, library, and natural history specimens. If on a sunny day, the trees surrounding the museum were removed and with the wave of a wand, the blue of the Mediterranean Sea became visible in the background, the building would look exactly like a minor Greek temple of great charm. The basically classic building has changed little over the years, except for modifications in the 1950s and again in 1981.

Inside, a gallery surrounds the two-story exhibit hall, and shelved cabinets line the east and west walls on both levels. Shaw built the museum to house his herbarium, library, and natural history specimens. It was a typical Victorian clutter of bird's nests, stuffed animals, and the collection of American birds from the Centennial Exposition held in Philadelphia in 1876. The interior shares a floor plan with Museum No. 2 at Kew, the Royal Botanic Gardens in London, a building that began as a place for fruit storage and was later converted to a museum.

When Shaw began his collection it is doubtful that he could foresee how the world would change over the next one hundred years. In Shaw's time the earth's surface was still being charted and explored; scientific optimism promised that no matter how difficult the problem, given world enough and time, a solution would be at hand. John Donne's notion that "No man is an island, entire of itself; every man is a piece of the continent, a part of the main . . ." was still just that, a

Opposite, above:
The Lehmann Building serves as the center of one of the world's largest programs of research in tropical botany. The library and the herbarium are housed within its reflective walls.

Opposite, below:
The Administration Building serves as offices for some of the Garden's staff. Much of the building originally stood in downtown St. Louis and was moved to its present site in 1890.

Ceiling detail of a mural painted on the ceiling of the Museum Building.

notion that referred to the philosophy of thought, having no relationship to the domain of science. What happened in the jungles of Brazil could have no connection with life in St. Louis or, for that fact, the civilized world.

The course of research has profoundly changed since the days of merely cataloguing ranks of stuffed animals as found in Shaw's original building. We now know that the actions of political entities can not only alter mankind's view of the pursuit of happiness but all the physical attributes of life itself. What happens in Brazil, to continue the example, could effect life as it was conducted and studied in St. Louis. Shaw's interest in science was part of the foundation that eventually lead to the Garden's emergence as a world leader in the scientific community.

THE FIRST FOUNDATION

During the winter of 1856, Shaw realized that there was much to learn about developing a large garden and wrote a letter seeking the advice of Sir William Jackson Hooker, the director of the Royal Botanic Gardens at Kew. Sir William, who as the first official director at Kew was the author of one hundred volumes devoted to systematic and economic botany, and well known for his kindnesses in assisting young botanists in their careers, had also established a museum of economic botany, a herbarium, and an extensive library during his tenure. He knew first-hand all the problems associated with developing such projects; luckily Sir William recognized Shaw's request as a serious one, and answered by sending descriptive catalogues of both Kew Gardens and the

Economic Museum, urging Shaw to remember that a great botanical garden is more than a fantastic collection of plants and sumptuous displays. Sir William also advised Shaw to contact Dr. George Engelmann, a St. Louis physician-botanist.

Meanwhile, Dr. Engelmann had written to Dr. Asa Gray of Harvard, who at that time was America's premier botanist, about Shaw, describing him as a "very rich Englishman . . . who [decided] to devote his whole time and fortune to the founding of a botanic garden and collection, Kew in miniature, I suppose." "Bravo Shaw!" wrote Gray, "I hope he will get a great many western things growing in his garden."

In 1856, Engelmann helped to organize the St. Louis Academy of Science, the first such institution to be established west of the Alleghenies. There he served as the middleman between Gray, at Harvard, and the plant explorers who were searching the West. Engelmann would send them instructions on collecting plants, later receiving the collected specimens either to be sold or sent on to Gray. Often the foragers would send seeds or drawings as well, and many of the plants turned out to be new species. Engelmann was also the first to call attention to the immunity of the American grape to the attack of insects called phylloxeras.

Gray was responsible for the *Manual of Botany,* a book that was found on the shelves of most identifiers of North American plants. Thus Henry Shaw had surrounded himself with three of the foremost botanically inclined minds of that time, all who agreed to help in the development of the Garden.

Scientific Beginnings

When Engelmann journeyed to England and Europe in 1857, he purchased the library and the herbarium of the German scientist, Johann Jakob Bernhardi, a collection that contained sixty thousand specimens that represented forty thousand different species, all at a cost of six hundred dollars. Bernhardi had been the editor of an important horticultural journal, and collectors from around the world had sent him samples of newly discovered plants, many which formed the basis for devising the botanical names for new species.

Now that Shaw had a museum, he needed a curator, and Gray recommended the hiring of Augustus Fendler, a gentleman who had recently settled in St. Louis and was one of the first plant collectors of America's Southwest. (Fendler is remembered as the finder of *Fendlera,* a genus of three shrub species native to Colorado and Mexico, more popular, ironically, in English gardens than American.)

Fendler has been described as the type of person happier roaming the mountains and deserts than getting dressed up in a suit and tie. He did arrange the botanical specimens, but after one year he gave in to wanderlust and left St. Louis. Unfortunately, the collection was left without guidance for the next thirty years.

The Civil War raged from 1861 to 1865, not without effect on Shaw's botanical oasis. Shaw continued to develop the Garden but the emphasis was not scientific. In 1867 he hired a fellow Englishman, James Gurney, as head gardener, and during the next twenty years the Garden became more popular and continued to garner rave reviews, although it was obvious to some, however, that both a competent curator and a general director were needed.

Dr. Engelmann died early in 1884. His son was willing to give all his father's specimens and personal library to the Garden, which would then put it second only to the collections at Harvard. But he also felt that Shaw would look upon Engelmann's collection as just more bundles of dried plants. Fortunately, the son was eventually persuaded to effect the transfer.

Meanwhile, Shaw had been asked to give the Garden to Washington University but Asa Gray advised against it. Gray knew from experience of the tendencies of schools and universities to subtly divert legacies to conform to their own wishes; instead he told Shaw to incorporate the Garden, appoint a Board of Trustees, and to itemize his concerns over botany, agriculture, and horticulture. Finally, Gray advised Shaw to endow a chair of botany at Washington University.

In June of 1885, the Henry Shaw School of Botany was founded, and Gray now advised Shaw to promote the hiring of Dr. William Trelease, a recent graduate of Harvard and at that time the professor of the Department of Botany at the University of Wisconsin, as the first Engelmann professor. Trelease was an excellent choice in many ways. He combined the disciplines of a scientific mind with a first-class sense of humor, and was quite willing to work for the success of the program.

Henry Shaw died on August 25, 1889. His will provides that the Garden should be "forever kept up and maintained for the cultivation and propagation of plants, flowers, fruit and forest trees, and other productions of the vegetable kingdom; and a museum and library connected therewith, and devoted to the same and to the science of Botany, Horticulture, and allied objects. . . ."

Early in September of that year the Garden trustees met to form a Board of Trustees, as stipulated in Shaw's will. Jointly they formulated rules and regulations for the Garden's operation. And they appointed Dr. William Trelease as the director of the Garden.

Trelease jumped in with both feet and early in January, 1890, presented a report on the progress of work being done to refurbish the Garden, which was then published in the first issue of the *Missouri Botanical Garden Annual Report.* The trustees also authorized the purchase of new botanical books at a cost of fifteen hundred dollars, the first book purchases since 1858.

The Garden truly entered the world of botanical exploration in 1890 when Trelease sent his first assistant, Albert Spear Hitchcock, on a natural history cruise of the Caribbean, marking the first time a member of the Garden's staff was directly involved in the collecting of plant specimens. Hitchcock later became the principal botanist of the

USDA Division of Plant Exploration and authored the definitive encyclopedia *Manual of the Grasses of the United States.*

Then in 1891, Trelease published a monograph on the fireweeds or willow herbs (*Epilobium*) of North America, a volume that became the first scientific publication of the Garden. That same year Trelease presented his personal collection of eleven thousand specimens representing four thousand species, plus his library of five hundred books and three thousand scientific pamphlets to the library.

In 1892, Dr. E. Lewis Sturtevant of Massachusetts, donated his collection of more than one thousand herbals and botanical books to the Shaw Library, including many pre-Linnaean volumes, those published before 1753, the date of publication of Linnaeus's *Species Plantarum.* When combined with the books from Engelmann and Trelease, the total collection of books became the foundation of one of the finest botanical libraries in the world. That year an inventory counted 11,455 books and pamphlets.

The course of research at the Garden was set.

PRESENT-DAY RESEARCH

The Missouri Botanical Garden is more than a garden. It is much more than rank upon rank of plants and flowers; more than a panorama of living color; and more than an educational facility to serve the residents of and visitors to St. Louis. The roots of the Garden reach throughout the Earth. They touch the other major botanical institutions and their related scientific communities in an expanding network devoted to the world of botanical research.

Today, scientific research at the Garden focuses on plant taxonomy—the describing, naming, and classification of plants—and systematics, the scientific discipline that formalizes this study. In addition to this general research, the Garden's scientists are involved with two major types of botanical work: monographic surveys that involve the study of one particular plant group, such as the entire poppy family, and floristic inventories where all the plants in a specific region are catalogued. To this end, Garden researchers are primarily involved with the exploration and study of the tropics, an area that includes the least known but most rapidly vanishing ecosystems of the planet.

Tropical evergreen forests cover seven percent of the world's surface, and two-thirds of all the plants and animals on Earth live only in the tropics. Of the two hundred and thirty-five thousand species of flowering plants in the world, half the total number of these plants occur in the New World tropics, south of the U.S.-Mexican border.

It is this plant-rich region that is drawing a great deal of critical attention as forests there are being burned deliberately to make way for so-called land development. What it does suggest is that the tropical forests are likely to be destroyed or permanently changed within the next fifty years. Presently there are worldwide discussions on the long-

The Garden was the first herbarium in the United States to use movable-aisle compactor storage units. Each of its 8 units consists of a central, stationary bay with seven movable bays on either side. These bays slide apart along tracks in the floor, forming 3-foot-wide aisles. Only one aisle may be open on either side of the bay at one time, thereby saving floor space. Within an aisle are 19 vertical rows, 14 shelves high on either side. Each of the Garden's large compactor units has the capacity of about a half-million specimens.

term effects of the slash-and-burn technique on the world's climate. Moreover, of what importance are the thousands of species that become extinct as these forests are destroyed?

Every day somewhere in the world, a child chews on a brightly colored berry from a plant collected in the woods or bites into the attractive fruit of a houseplant and is suddenly taken ill. Then a quick-thinking physician consults a book on wild flowers or pages through a handbook of poisonous plants, identifies the plant involved, and effects a cure. Someone else makes a cup of special herb tea to relieve a shattering headache; or a gardener becomes the envy of the neighborhood by planting a newly acquired annual in his border, a flower that outshines all others in the bed and unbeknown to the gardener, the result of an expedition to the deserts of China; and a scientist in a pharmaceutical house prepares a report that will eventually result in a new aid to arthritis. In every one of these cases, plant collection and plant identification is at the core of its success.

The entire history of the world revolves around plants: The grasses alone provide corn, wheat, rice, oats, barley, rye, sorghum, and sugar; thousands of medicinal products are provided by plants; the first paper in the world came from papyrus and today is manufactured from wood pulp; the basic materials used in the science of genetic engineering are contained in the DNA of wild plant species ranged throughout the world; gardens all over the world soothe both body and spirit; and for every plant that has been catalogued and examined for its benefits to a belabored humanity, there are thousands more yet to be discovered.

THE HERBARIUM

Botanists and scientists base their primary knowledge of plants on pressed and dried specimens that have been carefully labeled as to where and when they were originally collected, and then mounted, catalogued, and finally stored in dry, insect-free environments. Such a collection is called a herbarium.

With some three and one-half million specimens, the Missouri Botanical Garden herbarium is one of the largest in the United States and among the top twelve in the world. Plants from Africa, alone, now total five hundred thousand.

Given proper care, dried plants stored in a herbarium will last an extremely long time. The Garden's herbarium contains plants collected by Charles Darwin on his voyage with the HMS *Beagle* (the specimen is a fern, *Asplenium magellanicum*); plants once held by Captain James Cook on his first voyage in 1768; and plants originally from the collections of Bernhardi and Engelmann. When properly prepared such specimens retain the features needed to characterize, classify, and name them, even to the level of microscopic and chemical features.

And the herbarium continues to expand. Each year seventy-to-eighty thousand new plant specimens are added to the permanent collection, and the material usually includes a sample of the plant's flower, its fruit, and the leaves. Duplicate specimens are used to acquire additional material from other herbariums, as explained in the words of a Garden publication: "Much as youngsters trade baseball cards, exchanging [or loaning] cards already represented in their collections for cards they do not yet have . . . the Garden exchanges specimens with three hundred herbaria worldwide."

In order to facilitate both the cataloguing and eventual use of plant specimens for study, it is important that they be mounted in a way that permits handling.

The various specimens are glued to large sheets of paper—if glue is impractical, the parts are sewn to the sheet—then labeled in the lower right corner.

The Garden may loan out fifty thousand specimens each year to other herbariums so each sheet is stamped with its name and accession number.

When the plants arrive in the Garden's receiving room there is always the potential problem of insect stowaways. Here in St. Louis the only insect that presents a problem is the cigarette beetle, the pest of the tobacco industry. While packed tobacco leaves are aging in barrels, the beetles can chew through the top four or five inches of leaf; moreover, the beetles are threats to most dried plants so the herbarium must be kept sterile. The tobacco industry uses hormones that prevent the beetles from sexual activity; this way, they just keep eating and eating until they pop! As a prevention measure, the Garden now heat-sterilizes the plants in chambers at 150° for ten hours, killing all the life stages of the beetle. Since some botanists arrive with plants just for a few days' study, there is a microwave oven to handle smaller amounts.

The United States Department of Agriculture is continually worried about the importation of plant materials into the country, but their main concern is soil and the nematodes or other harmful soil organisms such shipments might contain. The plants sent to the Garden are carefully cleaned of any such soil.

Another consideration is the illicit trade in rare and endangered species, but a system exists to monitor this problem. Known as the Convention on International Trade in Endangered Species (CITES), this international commission publishes lists of species that are rare and endangered, for example, the entire orchid family is on the list. The botanical gardens of the world receive permits from CITES to move scientific material around for scientific study, displaying their CITES permit number and symbol on the packages.

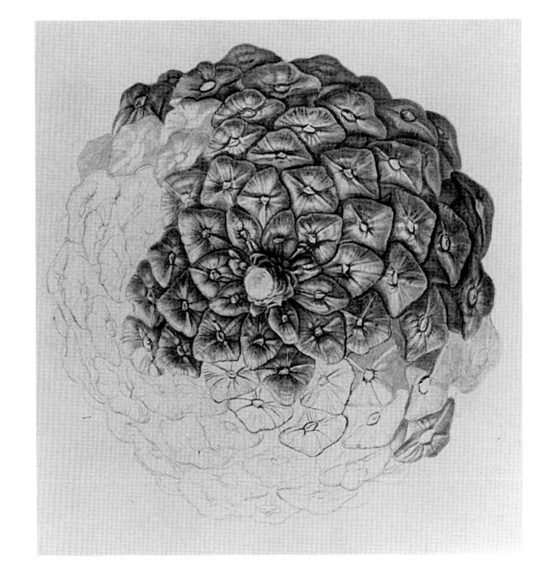

A drawing by P. Roetter of the end view of a pine cone taken from the George Engelmann Papers at the Library.

A dried specimen is positioned on mounting paper prior to being glued in place. It will then be filed in the herbarium collection.

THE LIBRARY

Most visitors to the library probably do not realize how much time has been taken to make the facilities so comfortable to work within. But it is the kind of place that begs to be used with its comfortable seating, pleasant levels of light, and, whether winter or summer, the beautiful views of the Garden visible through the large windows.

Discoveries in botany are rarely the result of fantastic finds. Rather they are the result of painstaking observation and research, conducted in the modern laboratory. Above, Dr. Peter Goldblatt, B. A. Krukoff Curator of African Botany, compares and examines dried specimens of the African butterfly iris, belonging to the genus Moraea. *Opposite, is an original watercolor of* Moraea loubseri *by Fay Anderson, taken from the book* Flowering Plants of Africa, *and at right, a dried and mounted specimen of the same species.*

Overleaf:
Display beds in front of the Lehmann Building. Three cockspur hawthorns, Crataegus crus-galli *'Crusader', are contained by plantings of bedding chrysanthemums.*

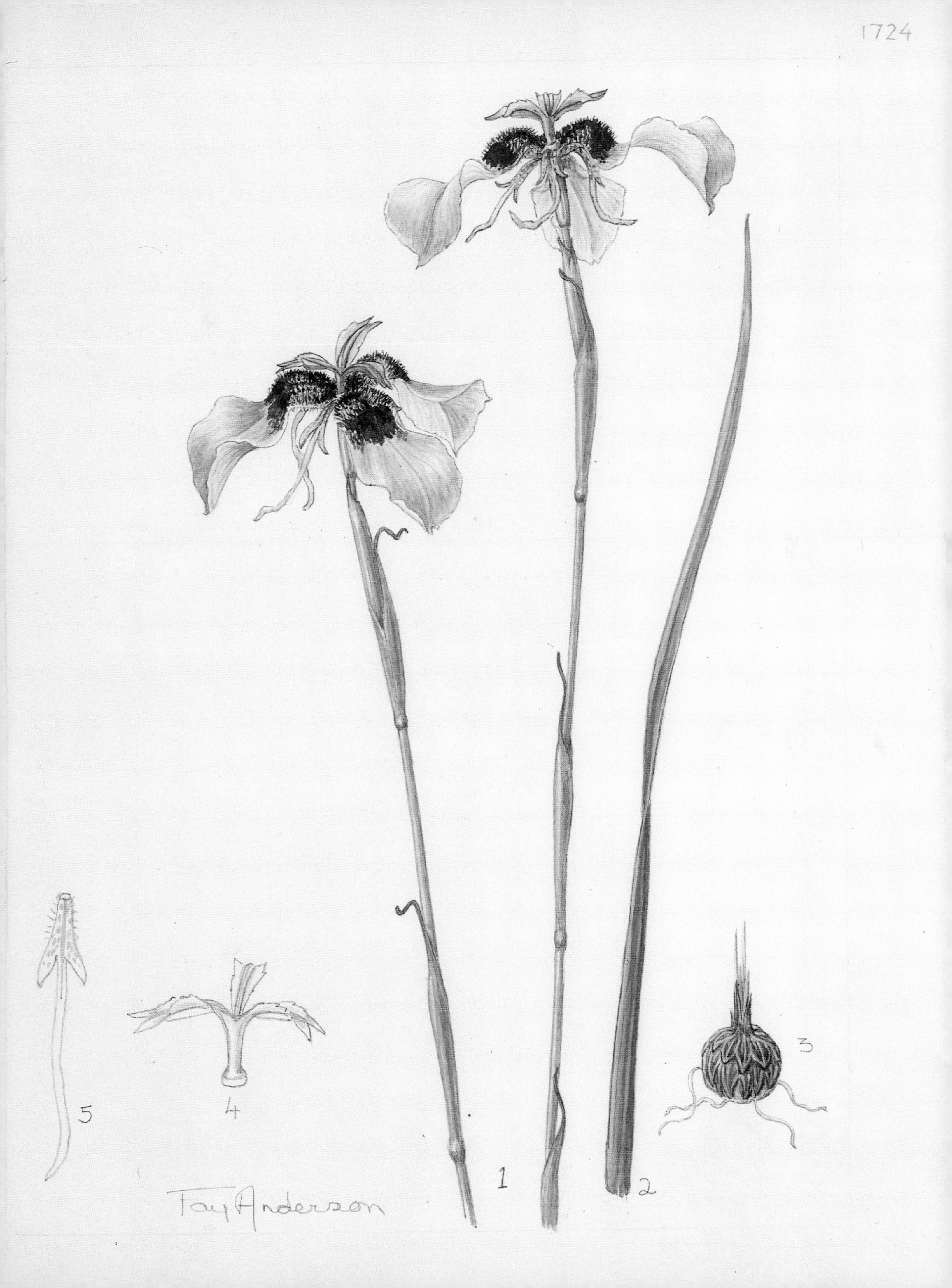

1724
1
2
3
4
5
Fay Anderson

The library is important not only for gathering together the news of research carried on today but offers help when it becomes important to learn what someone might have said years or centuries ago about a certain genus or species of plant. In absolute numbers—ninety thousand volumes—it is not a big library, but it contains a tremendously high proportion of the world's literature on botany going back to Linneaus and before. In addition to the general holdings, there is a rare book collection of over five thousand volumes with some books dating back to 1493. Extensive archival materials include more than two hundred and twenty thousand items, and the library receives thirteen hundred current periodicals.

COMPUTER RESOURCES

The Garden, like other huge enterprises, is now using a computer to store information on the plant collections. An ADDS Mentor 5000 computer is at the heart of the system. Routine herbarium management tasks such as the production of plant labels and keeping track of loan and exchange records are now completely computerized. Eventually it is hoped that a bar code system can be developed, much like the code that identifies products in a supermarket, to store information on each individual plant specimen.

The principal computer project now under way is TROPICOS, a master data base that started in 1983 and by mid-1988 contained over four hundred thousand entries of plant names, plant distributions, chromosome numbers, and related references to other botanical literature.

There are also probably one-quarter of a million known species of plants in the world with about one million names allotted to them. Another important computer job at the Garden will be removing all the extraneous duplicate plant names now in use.

LABORATORIES

Although much of the work connected with the scientific study of plants can usually be done with a microscope, a set of reference books, and a good eye, a well-equipped laboratory is maintained by the Garden for use by the research staff, students, and visiting scientists. The laboratory is particularly equipped for studies of cytology, the study of cellular makeup of plants, including the investigation and mapping of plant chromosomes.

Washington University has an electron microscope that is available to the Garden. Present investigations include studying the role of DNA in plant systematics research, in collaboration with researchers at the university.

Pots of seeds germinating in a working greenhouse.

Above, left:
The ground-floor level of a working greenhouse.

Over 250 herbs, shrubs, and trees belong to the genus Hibiscus. *Here the many stamens or pollen-bearing organs of a typical hibiscus flower are easily seen. This particular plant is a small tree,* H. waimae.

An experimental hybrid in the genus Centropogon *blossoms in the Research Greenhouse. In addition to aiding in the analysis of flower behavior, many such plants are used to study plant chromosomes and to develop new hybrids.*

Research associate James Aronson, working in north Chile, gathers the seeds of chañar, Geoffroea decorticans, *a legume tree found in arid and semiarid regions; the tree is useful for the firewood, fodder, and shade it provides. The seeds will be among those used in an experimental plantation and in studies of nitrogen-fixation in South American trees.*

Garden botanist Bruce Holst uses rock climbing equipment to explore for plants that inhabit hard-to-reach crevices on Aparamon Tepui in Venezuela.

GRADUATE TRAINING

Since 1885, when the Garden began its formal affiliation with Washington University, many influential plant biologists have received their graduate training as a result of this continuing partnership. Cooperative arrangements for graduate training have been established with other St. Louis area universities including St. Louis University, the University of Missouri-St. Louis, and Southern Illinois University at Edwardsville.

The Garden's director, Dr. Peter H. Raven, holds the position of Engelmann Professor of Botany at Washington University, and many Garden curators hold adjunct faculty appointments at one or more of the participating universities. Graduate students who are undertaking research for their doctoral or master's degree at the Garden will work under the direction of any of these regular or adjunct faculty members.

The outstanding programs and facilities offered by the participating universities—particularly in population biology and genetics, ecology, plant physiology, and molecular biology—used in combination with the facilities of the Garden and its research staff, lead to a particularly enlightened graduate program. Because of the Garden's strong commitment to tropical research and its field research programs in Latin America, Africa, and Asia, there are unique opportunities for field-oriented studies.

THE SCOPE OF WORLDWIDE RESEARCH AT THE GARDEN

The Garden now maintains one of the world's most active field research programs with a staff of approximately seventy researchers and technicians, including over thirty Ph.D.-level scientists.

Taped to the enameled side of one of the herbarium compactors is a copy of the large map of the world published by the National Geographic Society. Colored threads run from various countries and lead to groups of photographs that show the members of the Garden's exploratory staff and the various projects they are involved with. Although the major focus of field research is in tropical Latin America and Africa, scientists from the Garden have been members of expeditions to Israel, China, Malaysia, Japan, Australia, New Caledonia, New Zealand, and the Soviet Union.

Many threads run from Central America, for it is here that the Garden began its beneficent forays into the field of tropical botany. This narrow strip of land that connects North America proper and the Isthmus of Panama contains an extraordinary collection of examples of both northern and southern hemisphere plants.

The *Flora Mesoamericana* project, for example, represents an international collaborative effort between the Garden, the Universidad Nacional Autónoma de México, and the British Museum (Natural

On the front porch of the Hotel Tinalandia in Ecuador, field assistant Karyn Croat presses recently gathered plant material. Most of these specimens are aroids.

In Puerto Aparamon, Venezuela, Garden botanist Ronald Liesner presses plants between pages of local newspapers. After initial pressing, plants may be stored for weeks in alcohol in plastic bags until final preservation—using heat to dry them—is accomplished.

Far left:
Garden botanist, the late Julian Steyermark, prepares field notes and presses plants collected in the Venezuelan Andean state of Táchira. The notes form the basis for the information used in accompanying the plant specimens when they are filed in the Garden's herbarium in St. Louis.

Left:
In the Sierra de Juárez, Oaxaca, Mexico, Garden botanist Thomas B. Croat washes accumulated soil and organic matter from the root mat of a freshly collected aroid in preparation for field pressing. A portion of the stem will be kept alive so that observations on flowering can be made.

History) to catalogue all the plants of Middle America from southern Mexico to Panama.

In Nicaragua the Garden is involved with a comprehensive botanical inventory that will be published as a Spanish-language manual. During the ten-year period that saw the most intensive inventory, over eighty-five thousand individual collections were made. A similar Garden project will produce a field-oriented plant manual useful to the many biological researchers working in Costa Rica.

Then in Panama the Garden continues with the continual updating of its oldest tropical project, the monumental *Flora of Panama,* with its original enumeration of some sixty-five hundred species.

The threads from South America point to thousands of yet-to-be discovered plants growing within the vast and threatened rain forests, the expansive lowlands of the Amazon Basin, and the rain-drenched areas of coastal Colombia and Ecuador.

Colombia is host to the Garden for ten flora projects including a long-term study of the Chocó, one of the wettest and most biologically diverse regions on earth. And several other projects are underway, mostly in western and northern Colombia.

Venezuela is home to yet another monumental study, a forty-year inventory of the Guayana Highlands. In 1988 the Garden received a grant from the National Science Foundation to support Dr. Julian A. Steyermark's study entitled, "The Flora of the Venezuelan Guayana."

The Venezuelan Guayana is the site of Arthur Conan Doyle's 1912 adventure classic, *The Lost World,* where imaginary pterodactyls ruled the skies and dinosaurs roamed the jungle. A total number of nine thousand species is estimated to occur within this area, half of which cannot be found anywhere else in the world. Dr. Steyermark, who is credited with collecting over one hundred and thirty-eight thousand plant specimens over his lifetime, is listed in the *Guinness Book of World Records* as the "champion plant collector." He received his doctorate from Washington University and the Garden in 1933 and joined the staff in June, 1984, after a distinguished career at the Field Museum, Chicago, and the Instituto Botanicao, Caracas. Dr. Steyermark died in October, 1988, but the project will be completed by his associates at the Garden and the network of contributors that had been established before his death.

In Ecuador the Garden is presently conducting studies of that country's amazingly diverse orchid species and the preparation of a field guide to Amazonian tree species in order to provide Ecuador's forestry department with botanical information. This information is deemed essential for the implementation of proper forestry protection and management.

Peru is home to a botanical inventory (first begun in the 1920s) that now focuses on the rich Andean slopes and Amazon lowlands. Garden scientists have shown that the rain forests of the upper Amazon in Peru may be the world's richest source of tree species, with 283 species found in one 2½-acre patch of forest.

Recently, the National Geographic Society began supporting Dr. Henk van der Werff's field study of the plant family Lauraceae, the laurel family. Dr. van der Werff, assistant curator at the Garden, will continue to collect plant specimens of this imperfectly known family in the area around Iquitos in Amazonian Peru. The Lauraceae are poorly understood because most species are large trees with small, inconspicuous flowers that make collecting and identification difficult. The family is one of the largest and most important families of tropical woody plants and are the source of valuable timber, avocados, spices, and aromatic oils.

In Paraquay the Garden is involved with developing a system of national parks in the eastern portion of that country.

And in Bolivia, the Garden is sponsoring yet another floral inventory that currently concentrates on the wet Andean slopes of the Yungas region.

According to Dr. Marshall Crosby, assistant director of the Garden, the Garden's seeming concentration of collecting plants in tropical areas began almost accidentally in the 1920s when one of the staff botanists was in Colombia collecting orchids for display purposes.

"While wandering through the Canal Zone," said Crosby, "he met a retired gentleman who had a fantastic collection of orchids. The collection was turned over to the Garden's care, and the administration then rented space to set up a field station for bringing more orchids in from the wilds. Plants were cleaned, packed, and shipped back to St. Louis for the-then orchid house.

"More botanists started traveling down to Panama for collecting expeditions in the 1930s and the Garden embarked on an ambitious project of cataloguing and describing the entire flora of Panama; it now has the most complete collection of Panamanian plants in the world. When that program was nearing completion in the 1970s, we looked around for other places to explore and having this broad base of plant knowledge of Panama, we extended further into Central America and the Andes in tropical South America."

Research work also leads from Africa. Since 1970 the Garden has been the designated center for the study of African botany within the United States. It also works on a new project to provide wild-collected African plants to the National Cancer Institute for continued screening for new anti-cancer agents.

About forty-five years ago, Robert E. Woodson, Jr., then director of the herbarium, became interested in Africa and the Garden started a modest collection of plants from that country. In the late 1960s, B. A. Krukoff, a well-known tropical explorer and botanist, established an endowment fund at the Garden, with the proceeds to be used exclusively for the purchase of African plant collections. At that time, most of the purchases were made from missionaries, school teachers, and other amateur botanists working in Africa. Purchases continue from the same fund, though today sources are primarily from various African institutions. Later in the 1970s, Krukoff established a second endowment at the Garden, with the income now supporting a full-time curator of African botany.

In Tanzania the Garden works in cooperation with the World Wildlife Fund, conducting a botanical survey of the remaining forested areas with an eye to future conservation.

In Cameroon, Garden researchers explored both the rich lowlands and montane rain forests for new food plants, since only a few of the vast world of plants today are used as food. The rain forests of Cameroon are the center for the genus *Cola,* plants having attractive flowers that provide seeds used in flavoring soft drinks.

Madagascar is home to one of the most unusual and highly endangered island flora in the world. Here the Garden is cataloguing plants and helping to develop conservation programs.

Southern Africa is host to a continuing study involving both the mosses of the temperate and subtropical areas of southern Africa and the unique flora of the Cape region.

Finally, the Garden recently initiated a cooperative effort with the Chinese government involving the publication of a multivolume, synoptical flora of that country, covering an area more or less equivalent to that of the United States, but with a flora of perhaps thirty thousand species compared with twenty thousand for North America.

Back at home, the entire continent of North America is flagged for a number of projects, including a *Flora of North America,* a long-needed synopsis of the plants of the United States, Canada, and Greenland. For incredible as it sounds, there is no book published on the flora of North America. Five years ago the director of the Hunt Botanical Institute at Carnegie Mellon University suggested that the botanical community begin such a project, and today it is well on its way, with the organizational center at the Garden.

This amazing twelve-year collaborative effort will include the effort of some twenty United States and Canadian botanical institutions, and will end in the publication of twelve volumes covering the plants of North America along with a computerized data base. It is interesting to note that both the USSR and Europe have sponsored such publications.

Dr. Raven, the Garden's director, has received a grant from the National Science Foundation for a project entitled, "Systematic and Evolutionary Studies in the Plant Family Onagraceae." The grant supports a long-term study of the Onagraceae (evening primrose family), a family of almost seven hundred species with flowers as diverse as the popular *Fuchsia;* the sundrops and the evening primroses; the willow herbs; and the summer bedding annuals, *Clarkia,* or more popularly known as godetia. The objectives of the project are to complete a thorough systematic revision of the entire family, developing Onagraceae as a model for evolutionary studies, enabling the botanist to see the diverse development of this unusual plant family, all from a single ancestor.

And the Garden's work continues on a regional level, too. Early in 1988, the Garden announced that an updating of *Flora of Missouri,* by curator Dr. Julian Steyermark, was underway. Since Dr. Steyermark completed the initial publication in 1964, many new plants have been found in the state and many more botanical names and classifications have been changed. In order to facilitate the effort, George Yatskievych, a Missouri Department of Conservation botanist, has a full-time office at the Garden.

THE FUTURE OF RESEARCH AT THE GARDEN

About twenty years ago one of the botanists at the Garden was asked to search out and study some poppies reputed to grow in Iran, plants that were closely related to the opium poppy. There had been reports in the literature that these particular flowers contained a chemical that could be used in treating heroin addiction. The plants were found and the Garden conducted field studies to determine their taxonomy, followed by a long series of chemical evaluations. The results were less than encouraging: the side effects of this particular chemical were worse than the original addiction to the heroine. Even though this plant could not ultimately be employed to curb drug addiction, it is through the exhaustive research efforts at institutions like the Missouri Botanical Garden that this kind of important plant research can take place. The chances are slim for finding a so-called miracle cure for any disease. If after testing several thousand samples of various plant chemicals, only six wind up in clinical trials, scientists consider this a good average. The "success" of research is not solely whether the knowledge can be practically applied, but in the information gained from the research.

Dr. Enrique Forero is the director of research at the Garden. Considering the future direction of research programs, Dr. Forero projected an expansion of the Garden's enormous research legacy.

> Our efforts will consist primarily in the continued exploration of plants in the tropics in order to make the plants, their distributions, and their characteristics better known. In the future, our activities in tropical countries will relate increasingly to an understanding of the economic properties of plants and to their selection as important objects for cultivation and development. And we will continue to be concerned with environmental issues. Conservation attempts and sound management practices in tropical countries are faced with a lack of basic scientific information about the ecology and the biological diversity of the areas to be conserved or managed.
>
> The Garden has a responsibility to assist developing countries in their efforts to resolve these grave problems. Interactions between scientists from the Garden and the governments should have greater impact in the development of environmental policies.

And the marvelous thing about the Missouri Botanical Garden is that while involved in the continuing search for the new and the rare, they will continue to help in charting a sane environmental policy for all.

Overleaf:
Drum Bridge, Taikobashi, *in the Japanese Garden.*

INDEX